Walter N. Hartley

Air and Its Relations to Life

Being, with Some Additions, the Substance of a Course of Lectures Delivered in the

Summer of 1874 at the Royal Institution of Great Britain. Second Edition

Walter N. Hartley

Air and Its Relations to Life
Being, with Some Additions, the Substance of a Course of Lectures Delivered in the Summer of
1874 at the Royal Institution of Great Britain. Second Edition

ISBN/EAN: 9783337161934

Printed in Europe, USA, Canada, Australia, Japan

Cover: Foto ©berggeist007 / pixelio.de

More available books at **www.hansebooks.com**

AIR

AND

ITS RELATIONS TO LIFE

BEING, WITH SOME ADDITIONS, THE

SUBSTANCE OF A COURSE OF LECTURES DELIVERED
IN THE SUMMER OF 1874 AT THE ROYAL
INSTITUTION OF GREAT BRITAIN

BY

WALTER NOEL HARTLEY, F.C.S.

DEMONSTRATOR OF CHEMISTRY AND LECTURER ON CHEMISTRY IN THE
EVENING CLASS DEPARTMENT, KING'S COLLEGE, LONDON

SECOND EDITION

LONDON
LONGMANS, GREEN, AND CO.
1876

PREFACE

TO

THE SECOND EDITION.

———◦◦◦———

THE REPRINTING of this little work was required before I had prepared for it, and during those months when my duties are particularly heavy. Nevertheless Prof. Tyndall's recent experiments have been described, and certain errors have received correction. I have been much aided in the revision of the book by Mr. CHARLES PULLER, lately a Fellow of Trinity College, Cambridge, to whom my cordial thanks are due for the care and interest he has taken in giving me help.

KING'S COLLEGE, LONDON : *May* 2, 1876.

PREFACE

THE FIRST EDITION.

———◆———

IN the following pages I have endeavoured to give, in a light and popular manner, some information concerning that particular form of matter called Air, which is so essential to man that it comes to each individual with life, and leaves him not till death. Besides the narration of facts, an account of how these facts were obtained offers an insight into the particular mode of reasoning employed in scientific research, and endows the statements with that weight and interest necessary to leave a distinct impression upon the mind. As far as is consistent with clearness of expression, the use of scientific terms has been avoided. In one or

two cases important practical details have been
more fully treated than usual. Generally speak-
ing, the original sources of information have been
consulted, and to some readers it may be of
interest to know that these consisted of the
papers of Andrews and Tait, of Graham, Ray
Lankester, and others, in the ' Philosophical
Transactions,' and in the ' Proceedings of the
Royal Society'; likewise the contributions of
Laplace, Dumas, and Boussingault, Regnault,
Lewy, Schoenbein and Pasteur, in the 'Annales
de Chimie et de Physique' and the 'Comptes
Rendus' of the French Academy ; of Bunsen,
Brunner, and Pettenkofer, in the 'Annalen der
Chemie,' &c. &c.   I am further indebted to
Dr. Roscoe's article ' Air,' in ' Watts' Dictionary';
Dr. Angus Smith's ' Life of Dalton '; also his
very important work on ' Air and Rain '; ' Prac-
tical Hygiene,' by Dr. Parkes ; ' Hétérogénie,' by
M. Pouchet ; De Bary's ' Morphologie und Phy-
siologie der Pilze.'; Cohn's ' Beiträge zur Biologie
der Pflanzen'; ' Fungi,' by the Rev. M. J. Berkeley,
in Hooker's ' English Flora '; and Dr. M. C.

Cooke's recently published work; also Pettenkofer's 'Popular Lectures,' translated by Dr.
Augustus Hess; besides many other writings,
to enumerate which would furnish too long a list.
Those who are acquainted with Prof. Bloxam's
work on 'Chemistry' will recognise some of the
experimental illustrations and wood-cuts employed. I have to acknowledge that my constant
intercourse and friendship with him have given
me advantages of which I have availed myself.
New illustrations, through the liberality of Messrs.
Longmans, have been engraved by Mr. Collings
from my own drawings, but for some of those
in Chapter IV. my thanks are due to the Council
of the Royal Society. The favour accorded to
the short course of lectures which I had the
honour of delivering at the Royal Institution of
Great Britain, in the summer of 1874, and the
subsequent enquiries as to their issue in a printed
form, has been the chief inducement to their
rearrangement and publication. Furthermore,
having been struck with the very general want of
knowledge in this country, even among scientific

men, of the work accomplished by that celebrated French chemist, M. Pasteur, I thought that to make his labours more widely known would greatly promote scientific truth and accuracy.

For the sake of accuracy it may as well be stated here, with reference to a sentence on p. 1, that although the rotation of the earth is said in general language to be accomplished once in 24 hours, it is really completed in 23h. 56m. 4s., and that the time of its actual revolution round the sun, the duration of a sidereal year, is 365d. 6h. 9m. 9·6s.

WEST DULWICH : *September* 1875.

CONTENTS.

CHAPTER I.

CHAPTER II.

CHAPTER III.

CHAPTER IV.

CHAPTER V.

LIST OF ILLUSTRATIONS.

AIR

AND

ITS RELATIONS TO LIFE.

———•◦•———

CHAPTER I.

Proofs of a Material Medium surrounding the Globe—Galileo's Theory, and Torricelli's Proof of its Weight, in the Year 1640—Priestley's Discovery of Oxygen, and Lavoisier's Discovery of the Nature of Air, in 1774—The Composition of Air and the Properties of its chief constituents demonstrated—Dumas and Boussingault's exact Experiments in 1841—Regnault, Bunsen, and Lewy's Determination of the Oxygen—Air from different parts of the World—Conclusions as to its Composition—Later Researches of Dr. Angus Smith—Reasons for considering Air a Mixture, and not a Compound, of Oxygen and Nitrogen—Tessié du Motay's Process for extracting Oxygen from the Air.

ASTRONOMERS tell us that the Earth has the figure of an oblate spheroid, that it revolves on its axis once in 24 hours, and traverses its orbit in 365¼ days; also that its shortest diameter measures 7927·4 miles. There is no occasion to question this statement, yet it may be suggested that this latter measurement is true only so far as the visible

B

portion of the earth is concerned. A correction
amounting to the addition of *at least* 90 miles
to this number must be made on account of the
surrounding invisible gaseous envelope or shell,
which, by its action on the sun's light, has been
computed to be not less than 45, and is probably
200, miles in thickness. It moves in the same
orbit and on the same axis, and is in reality a most
essential part of the globe, and more especially so
to all living creatures. This portion is called the
atmosphere. We are cognisant that we inhale this
atmosphere, we are aware that it affords a certain
resistance to the sails of a boat, that something
preventing the entrance of water fills a glass or
bottle plunged mouth-downwards below its surface,
that this something has no definite shape, is without colour, taste, or odour, and is altogether unlike
the other forms of matter around us.

We know that air is essential to life, yet it is
more by being so taught than by intuition that we
have this knowledge, for, as Dr. Angus Smith
remarks, 'When we are children air is to us nothing ;' 'a vessel of air is a vessel with nothing in
it.' Air enters our bodies by the lungs ; a portion
is absorbed by the blood, and courses through the
system ; we cease to exist without it, and as living
creatures cannot rise above it ; it is at one and the
same time a part of the earth, and a part of nearly

all living things, animal and vegetable, upon the
earth. It constitutes by far the greatest portion of
the atmosphere ; so much so, indeed, that the words
air and atmosphere may be looked upon in most
cases as having the same meaning. A history;
therefore, of so important a matter as air cannot be
otherwise than interesting to many readers.

That the atmosphere was a material substance
was believed by Aristotle, but he did not succeed
in establishing the truth of his convictions. The
earliest experiment which failed to yield the desired proof may easily be repeated by anyone, and
may be discussed as follows :—

All material substances possess weight and
occupy space ; if, therefore, we take a bladder of
air and weigh it, then extract the air and weigh
it again, it might at first sight appear that the
second weight would be less than the first by
the weight of the air abstracted ; but under ordinary conditions of *weighing in air* it is not so,
because the experiment as thus performed, as far
as the object in view is considered, gives us nothing more than the simple weight of the bladder
itself. To make this more apparent, the experiment may be modified by substituting water for
air. Fill a bladder with water, sink it in a vessel
of water, and, by means of a string, attach it to a
pair of scales ; weigh it ; now, without removing

the bladder from below the surface of the liquid, squeeze it until it is as nearly as possible empty, tie up the mouth, and weigh again ; the results of the first and second weighings will be identical. The conditions of the two experiments are analogous. In the former case a bladder first full of air and then empty is weighed in air, in the latter a bladder of water, first full and then empty, is weighed in water, the conclusion in each case being the same. The reason why no satisfactory proof is thus obtained is accounted for by the following explanation. When a substance is weighed during immersion in any gaseous or liquid medium, the weight observed is smaller than the true weight by the weight of the surrounding medium displaced by the substance. A cubic inch of gold, for example, if weighed in air gives not its true weight, but a quantity which, if added to the weight of a cubic inch of air, is equal to its true weight. Or the same thing may be stated thus : if a cubic inch of gold be weighed in water, it gives its true weight minus the weight of a cubic inch of water. A bladderful of air weighed in air therefore yields a quantity which is minus the weight of the air occupied by the distended bladder, and these two weights are practically identical.

The apparatus with which the true weight of air was discovered consists of a glass globe of about a quart capacity, fitted with an accurate stopcock, the air being entirely withdrawn from the flask by means of an air-pump, and the vessel suspended and accurately counterpoised on a balance. It is evident that if the tap be now opened the air will rush in, and this quantity of air being sufficiently heavy to turn a good balance, the beam descends on that side sustaining the flask.

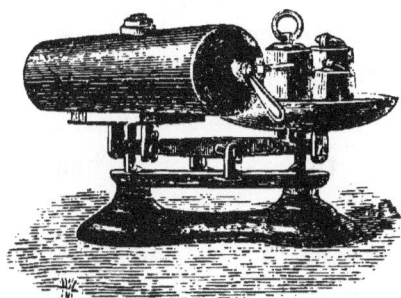

Fig. 1. Method of showing the Weight of Air.

One decisive experiment to show that air has weight may, as it is a novel one, be mentioned here. An iron bottle, into which $7\frac{1}{2}$ cubic feet of air have been pumped, is exactly balanced on a pair of scales ; by allowing the air to escape, and, weighing again, it may be found to lose more than 7 ounces, and weights to this amount must be placed on the cylinder to restore equilibrium (see fig. 1). As the air flows into an empty gas bag it is shown to occupy space. What further proof is necessary? Air is evidently a material substance. At a temperature of 60° F., the height of the barometer being 30 inches, 13 cubic feet of air would weigh one pound. An experiment done on so large a scale and with such rough weighing apparatus as the above may be considered to yield a very fair indication of this fact.

Galileo gave rise to a great discovery when he first explained to the pump-makers in Florence that the weight of a column of the atmosphere was not sufficiently heavy to balance a similar column of water more than 32 feet in height, hence the reason why they had failed to raise water in pumps to a greater height than 32 feet. Torricelli, his pupil, argued that the atmosphere which would support a column of water 32 feet would support a similar column of mercury only 30 inches in height, because mercury is 14 times heavier than water. This was verified by the following experiment, and so the first barometer was constructed :—A tube about 33 inches in length is filled to the top with mercury; being closed with the finger, it is inverted, the closed end dipped below mercury contained in a dish, and the finger removed ; the mercury is then found to sink a little more or less than 3 inches. Now the weight of a column of mercury 30 inches in height and with a surface of

1 square inch would be 15 pounds ; the weight, therefore, of a column of the whole 45 miles of atmosphere an inch square would be 15 pounds, because it exactly balances 15 pounds of mercury. Hence the barometric pressure is said to be 15 pounds on the square inch.

The necessary consequences of the weight of the air which surrounds all terrestrial things are curious. For instance, the temperature at which water boils is 100° Centigrade, or 212° Fahrenheit, at the sea-level. If we decrease the pressure on the surface of the water we lower the boiling point, and this may be done by placing it under an air-pump and extracting some of the air surrounding it, when it can be very readily made to boil at the temperature of a warm summer day. Or we may lessen the pressure on the surface by ascending a high mountain, and so diminish the amount of atmosphere which rests its weight on the liquid. This Professor Tyndall has done, and he states that the boiling point of water on the top of Mont Blanc is 184·95° F., or 27° below the boiling point at the sea-level. This difference is due to the removal of the pressure caused by a layer of air 16,000 feet in height. A variation of 1° F. indicates an ascent of 596 feet. It follows from this that if we increase the pressure the boiling point will be raised ; and it has been found by Regnault that at twice the atmospheric pressure, or 30 pounds upon the square inch of surface, water boils at 249·5° F.—a rise in temperature of 37·5° F. Were it not for the weight of the atmosphere the ocean would evaporate, all living things, vegetables, and animals could no longer exist, and the earth would roll on its course a dull lifeless rock, clad here and there with ice, and surrounded by aqueous vapour.

The superincumbent mass of the atmosphere, which exerts a pressure expressed by the term 15 pounds on the square inch, but the extent of which may be made more apparent to you by the statement that it amounts to 14 or 15 tons on the body of a man of average size, has its counterpart in the enormous pressure of water at great depths of the ocean. The dredging apparatus on board H.M.S. 'Challenger' has brought up fishes from great depths, distended and deformed, with their swimming bladders protruding from their mouths, because on coming to the surface the relief of the great pressure under which they were accustomed to live caused the air within them to be greatly expanded. Thermometers let down, which had been submitted to a pressure of 4 tons, were broken simply by the weight of the superincumbent water being greater than this. The iron cylinders filled with compressed gases, such as may be used in the foregoing experiment, usually sustain a pressure of 30 atmospheres, or 30 x 15 = 450 pounds on the square inch.

It is easily seen, then, that our condition on the earth is that of creatures living at the bottom of a vast ocean which floods the entire globe.

This gaseous ocean has its tides which ebb and flow from exactly the same causes as those of the aqueous ocean. Laplace was the first to draw attention to this in his ' Mécanique Céleste.' These atmospheric tides are due to three causes— first, the direct action of the sun and moon upon the atmosphere; second, the periodical elevation and depression of the ocean; third, the attraction of the atmosphere by the ocean, the form of which changes periodically. These three causes are all derived from the attractive force of the sun and the moon, consequently the atmospheric and oceanic tides occur at the same periods. The atmospheric like the oceanic flow is due to the combined action of two partial tides, one caused by the sun, the other by the moon. We have evidence of the tidal flow of the atmosphere in the diurnal variations of the barometer. Extremely accurate observations, made every day for six con- secutive years, showed that the mercury is always highest at nine o'clock in the morning and lowest at three in the afternoon; and at Paris the differ- ence between the two readings amounts to eight-

tenths of a millimètre. Thus far we have had to
deal with such of the physical properties of the
atmosphere as are common to many other material
substances ; but in regarding it from a chemical
point of view we must submit it to a closer examination, learn with what special properties it is
endowed which distinguish it from all other kinds
of matter, and likewise what are its relations to

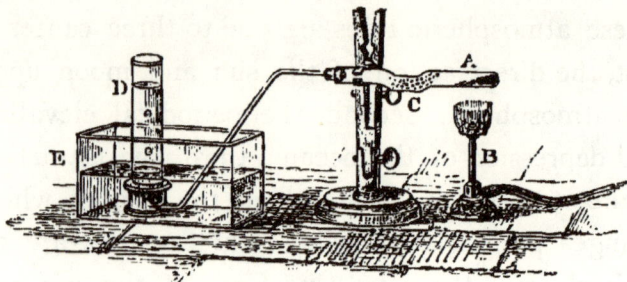

Fig. 2.   Preparation of Oxygen from Red Precipitate
or Oxide of Mercury.

A Oxide of mercury in a glass tube.
B Gas-burner.
C Metallic mercury.
D Cylinder of water inverted in the pneumatic trough E to collect
he gas.

other substances, and more particularly to the
various forms of living things.

Rutherford in 1772, in Edinburgh, discovered
an inactive gas as a constituent of the air, but the
discovery by Priestley in August 1774, a century
since, that when he heated red precipitate (see fig.
2) he obtained metallic mercury and a gas or air

of remarkably active properties, which caused a glowing taper to be rekindled and combustible bodies to burn with increased brilliancy, paved the way to the discovery of the nature of air by Lavoisier still later in the same memorable year 1774. The experimental proof employed by the French chemist consisted in the calcination or burning of mercury at a moderate temperature

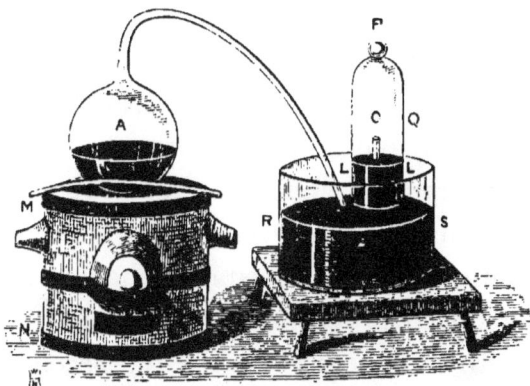

Fig. 3.   Lavoisier's Experiment.

and its revivification by increased heat with the simultaneous production of oxygen gas. A translation of Lavoisier's own words affords the best description of his experiment:—

'I took a flask of about 36 cubic inches capacity, the neck of which was very long, with an internal diameter of half an inch. I bent it in the manner seen in fig. 3, so that it could be placed on the furnace M N, whilst the extremity O of its neck

fitted under the bell-glass P Q, placed in a bath of
mercury R S. I put into this flask 4 ounces of
very pure mercury ; then by sucking with a syphon
introduced into the bell-glass P Q, I raised the
mercury to L L. I carefully marked the height of
this with a band of paper, and observed exactly
the barometer and thermometer. These prepara-
tions made, I lighted the fire in the furnace M N,
and sustained it for very nearly twelve days in
such a way that the mercury was heated almost to
its boiling point. Nothing remarkable occurred
during the first day. The mercury, although not
boiling, was in a continual state of evaporation ; it
coated the inside of the vessel with globules, at first
very minute, which increased until, having obtained
a certain size, they fell back into the vessel and re-
united with the mercury. On the second day I
began to see little red particles swimming on the
surface, which for four or five days continued to
increase in number and size, after which they
ceased to grow and remained in absolutely the
same state. At the end of twelve days, seeing
that the calcination of the mercury (oxidation of
the mercury) made no further progress, I withdrew
the fire and left the vessels to cool. The volume
of air contained in the flask, its neck, and the bell-
glass, reduced to a pressure of 28 inches and a
temperature of 19° Centigrade, was before the

operation about fifty cubic inches.[1] When the operation was concluded the volume, under the same conditions of temperature and pressure, was found to be no more than forty-two or forty-three inches; there had consequently been a diminution in volume of about one-sixth. On the other hand, having collected carefully the red particles which were formed, and separated them as much as possible from the mercury with which they were mixed, their weight was found to be forty-five grains.

'The air which remained after this operation, and which had been reduced to five-sixths of its original volume by the calcination of mercury, was no longer fit for respiration or combustion; animals placed in it perished in a few moments, and lights were extinguished immediately, as if they had been plunged under water. On the other hand, I took the forty-five grains of red matter formed during the operation, and I introduced them into a little glass retort which was fitted to an apparatus meant to receive the liquid and gaseous products that might separate. Having lighted the fire in the furnace I observed that as the red substance was heated its colour increased in intensity. When the

---

[1] As the space occupied by a gas, that is to say, its bulk or volume, varies with the temperature and the barometric pressure, these details are necessary for exact measurement.

retort became nearly red-hot, the red substance
began gradually to diminish in quantity, and after
some minutes it had disappeared altogether ; at
the same time there were collected in the receiver
41½ grains of mercury, and there passed into the
bell-glass 7 to 8 cubic inches of an elastic fluid
much more fit than atmospheric air to support
combustion and the respiration of animals. Having
passed a portion of this gas into a tube of glass
an inch in diameter, a taper was plunged into it
and diffused a dazzling brilliancy ; the carbon, instead of burning quietly as in ordinary air, burnt
with a flame and a kind of decrepitation like that
of phosphorus, and an intensity of light that the
eyes could scarcely support.

'On reflecting on this experiment, it is seen
that mercury during calcination (oxidation) absorbs the wholesome and respirable portion of the
air ; that the portion of the air that remains is a
kind of choke-damp, incapable of supporting combustion and respiration.   Atmospheric air, then, is
composed of two elastic fluids of a different and, so
to speak, opposite nature.

'A proof of this important truth is, that in re-
combining the two elastic fluids which have thus
been obtained separately, that is to say, the 42
inches of choke-damp or irrespirable air, and the
8 cubic inches of respirable air, air is re-formed in

every particular like that of the atmosphere, and which is equally fit for combustion, the calcination of metals, or the respiration of animals.'

Lavoisier adds that the proportion of respirable gas found by his experiment is probably a little too small, because he could not get the entire quantity of the constituent oxygen to combine with the mercury.

The conclusions of Lavoisier have been verified and corrected by later experiments more easily performed ; but the great neatness of his demonstration, together with the light it throws on the action of air in the phenomena of combustion, casts a glory over the whole investigation.

That air consists for the most part of 4 volumes of nitrogen and 1 volume of oxygen we show now-a-days in various manners, for instance, by suspending a stick of phosphorus upon a wire stand in a measured volume of air confined over water in a glass cylinder. The glass cylinder is divided into five equal parts by measuring water into it, and each space is marked on the outside by black lines.

After a few hours the phosphorus, having combined with all the oxygen to form phosphorous acid, a substance which is absorbed by the water, leaves the nitrogen oocupying only four of the spaces marked on the jar.

We may illustrate the same fact in a much

more striking way, by the following new method (see fig. 4) :—A good glass tube closed at one end, having an internal diameter of 1 inch and a length of 45 inches, is fitted with an india-rubber bung. The tube should then be divided into five equal parts from end to end by small india-rubber rings stretched over it. A perfectly dry piece of phosphorus is placed within the tube about 3 inches from the india-rubber stopper; this is warmed until well melted, and then jerked down to the other end of the tube. In its progress it takes fire and so uses up all the oxygen. The tube soon becomes sufficiently cool for the end to be dipped beneath coloured water, when on removal of the cork the

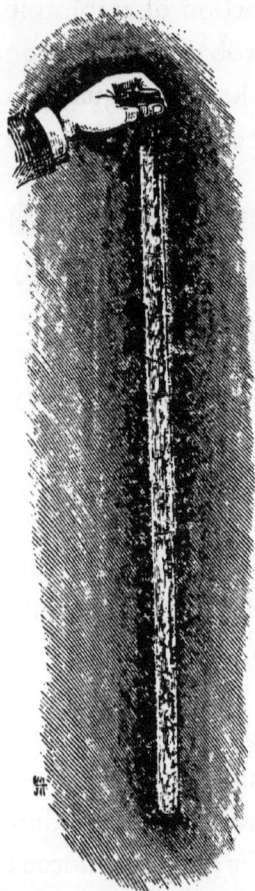

Fig. 4. Experiment to show the Composition of the Air.

liquid rushes in to supply the place of the missing oxygen. On raising or depressing the tube so as to bring the liquid within and without to the same level, inserting the cork again beneath the water and removing, it is seen that the liquid within occupies one-fifth the entire capacity of the tube, which if the tube be of equal bore throughout will be as nearly as possible 9 inches. See fig. 4.

It having thus been demonstrated that there are two bodies, nitrogen and oxygen, present in the air, let us examine their properties. First the nitrogen. Taking a bottle full of this substance it is seen to be a colourless gas, for the bottle has the appearance which we generally term empty. Removing the stopper there is no smell detected, and placing a taper within, it is found not to be an empty bottle, for the taper is extinguished; another is lighted, and this is extinguished; the gas is therefore neither combustible nor a supporter of combustion. On placing some water in the bottle and shaking it up, the mouth being closed by the hand, we do not find the water altered at all; it is therefore not soluble to any great extent. All positive properties are 'conspicuous by their absence.' It is colourless, odourless, tasteless, incombustible, irrespirable, insoluble, and a non-supporter of combustion. When, however, it is combined, on the one hand, with hydrogen to form ammonia, or with oxygen and hydrogen to form nitric acid, we have in either case a new substance produced which is remarkable for its most strongly pronounced

C

properties, and is of the greatest chemical activity ;
the strangest fact is that these two substances,
ammonia and nitric acid, have exactly opposite
properties, but on mixing they combine together
and neutralise the distinctive characters of each
other.

With regard to the other constituent, oxygen,
all the positive character of air is due to it.   In

Fig. 5.   Charcoal burning in
Oxygen.

Fig. 6.   Phosphorus burning in
Oxygen.

other words, the properties of air are the properties
of oxygen much enfeebled.   We therefore find that
all substances which burn in air will burn with
greater brilliancy in oxygen, while some, such as
iron and zinc, which are incombustible in air, are
readily consumed in oxygen.

Experiments arranged to show the burning in

oxygen of charcoal, phosphorus, and also a steel
spring, are shown in the accompanying illustrations
(see figs. 5, 6, and 7). In the first case it is difficult to get the charcoal to take fire at all in air,
but having at last kindled it so as to get a minute
point to glow, on placing it in the oxygen it bursts
into a brilliant glittering shower of sparks. With
phosphorus the scarcely luminous and ghost-like

Fig. 7. A Steel Spring burning in Oxygen.

flame becomes a surprisingly brilliant blaze, and
even a steel watch-spring is consumed like a piece
of charcoal. The opposite characters and the
different properties of the two gases are thus
forcibly demonstrated.

The question would arise in the mind of every
chemist after repeating the experiments of Lavoisier
—What is the constitution of the atmosphere: are

these two gaseous constituents simply mixed together, or are they chemically combined? Are
the oxygen and nitrogen combined in the proportion of 1 volume of the former to 4 of the latter, so
that they produce a third distinct substance called
air, just as sulphur and quicksilver can be combined
together to form a third substance, vermilion; or
are the two gases merely mixed? While many
chemists, amongst whom we find the names of
Döbereiner,[1] Prout, and Thomson, considered, not
without good reason, the atmosphere to be a compound of 20 volumes of oxygen and 80 volumes of
nitrogen (that is 1 to 4), Dalton, on the other
hand, supposed it to be a mixture differing in composition at different altitudes according to the relative weights of the two gases, but this conviction
was founded less on experiment than calculation.
At first it appeared probable that the unequal
development of animal life on various parts of the
globe would effect considerable alteration in the
air of different localities, and this idea was supported not only by the unhealthiness of the air of
certain districts, such as crowded towns and marshy
lands, but also by the earliest quantitative analyses,
which showed differences in the composition of air
of different localities amounting to 10 per cent.

---

[1] The professor under whom Goethe studied chemistry at
Goettingen.

Babinet, indeed, went so far as to calculate from the air of Paris the probable composition of the air at various heights up to an elevation of 10,000 mètres, or more than 32,500 feet. Dalton, Gay-Lussac, Davy, and Boussingault, however, by making use of improved methods of analysis, found the oxygen to vary but slightly. Gay-Lussac and Thénard, in a memorable balloon ascent, at an altitude of 7,000 mètres, obtained a sample of air which gave the same numbers on analysis, namely 21·6 per cent., as a sample taken at the surface of the earth in Paris at the same time. This proved that altitude does not affect the relative proportion of the two constituent gases in air. This was afterwards confirmed by Brunner, who analysed air collected on the summit of the Faulhorn.

But it was necessary to have more accurate methods of analysis than those hitherto employed.

The experiments of Dumas and Boussingault in November 1841 finally established with great exactitude the true composition of air. Their apparatus (see fig. 8) consisted of a tube C C, provided with stopcocks, and containing metallic copper. The whole being weighed previously to being heated to redness in a furnace was connected at the one end with a series of tubes A, containing potash to absorb carbonic acid, with another set B, containing sulphuric acid to absorb ammonia and water, and at the other end with a large glass globe N, provided with a stopcock, exhausted of air, and accurately weighed. On slowly opening the taps the pressure of the external air caused it to flow through the series of tubes, and becoming perfectly purified, to gain admission to the heated metallic copper; here the oxygen was retained by the copper, forming a black solid metallic oxide, while pure nitrogen filled the globe. All the stopcocks were then

Fig. 8. Apparatus for the exact Analysis of Air.

closed, and the globe was detached and weighed: the difference between the previous and this second weighing expressed the weight of the nitrogen which had entered the globe. The tube was then separated and weighed a second time, then exhausted at the air-pump and weighed a third time: the difference between the second and third weighings gave the weight of the remnant of nitrogen in the tube, that portion not capable of passing into the globe. The difference between the first and second weighings of the tube expressed the weight of oxygen held in combination with the copper.

For example, the figures representing the result of the analysis are here given :—

Weight of		In Grains
Globe (N) with nitrogen (at the conclusion).	. .	. 30˙6
Exhausted globe (at the commencement) .	. .	. 3000
Nitrogen in the globe ✔.	. . . .	. 76
Tube (C) with residual nitrogen (at the conclusion)	.	. 2574
Exhausted tube (at the conclusion) .	. . .	. 2573
Nitrogen remaining in the tube	. . . .	1
Nitrogen in the globe .	. . . . .	76
Nitrogen remaining in the tube .	. . . .	1
Total nitrogen in the air analysed .	.	77
Exhausted tube (C) with oxidised copper (at the conclusion)		2573
,, ,, metallic copper (at the commencement)	. . . .	. 2550
Oxygen in the air analysed	. . . . .	23

The air therefore contained 23 parts of oxygen by weight and 77 parts of nitrogen.

The results obtained led to the conclusion that air, purified from water, carbonic acid, and ammonia, had the composition by weight of 77 parts of nitrogen and 23 parts of oxygen.

Analyses on three successive days, viz. :

Proportion of Oxygen in 100 *parts of Air.*

1841	By Weight		By Volume
	1st Experiment	2nd Experiment	
April 27. .	. 22˙93	22˙93	20˙729
,, 28 .	. 23˙03	23˙09	20˙828
,, 29 .	. 23˙03	23˙04	20˙828

It will be noticed in these numbers of Dumas and Boussingault that there are slight differences on different days in the proportion of oxygen present in 100 parts of air. With such a cumbrous apparatus and so elaborate a method of experimenting, it was not possible to make any very great number of experiments, but by the later method of gas analysis pursued by Bunsen, Regnault, and Lewy thousands of experiments have been made. This method depends on the fact that oxygen and hydrogen gases combine in the proportion of one volume of the former to two of the latter gas to form an inappreciably small volume of liquid water. By decomposing water with a galvanic battery, and collecting the two gases separately in a vessel made for the purpose (fig. 9), it is easily seen that the hydrogen gas in the tube *h* is double the volume of the oxygen in *o*. That the gases when mixed and

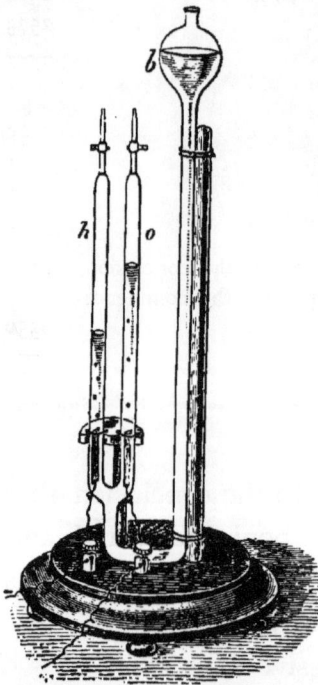

Fig. 9.   Apparatus for the Decomposition of Water.

then combined entirely disappear, may be made
obvious to the eye by exploding the mixture over
mercury in the following manner:—The mixed
gases are obtained from acidified water, which
is decomposed in a voltameter (A, fig. 10) by means
of a galvanic battery. The platinum plates B
(the essential parts of the voltameter) are attached
to the opposite poles of the battery by the wires C
and D, and the gas is collected in the stout glass
tube E, which has been previously filled with water.

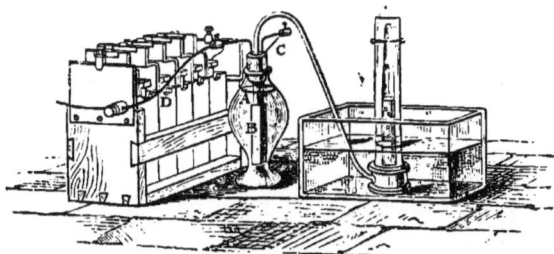

Fig. 10. Gases resulting from the Decomposition of Water being
collected in a Glass Tube.

To conduct the electric spark to the interior of the
tube two stout platinum wires are cemented into
holes drilled through the glass near the closed end.
When the tube is filled with the mixed gas it is
closed with the thumb and removed to a thick
porcelain mortar containing mercury, where it is
firmly pressed down on a cushion of stout vul-
canised india-rubber; the spark is now passed
through the gas from one wire to the other by

means of an induction coil, an explosion takes
place with a strong concussion but no noise, and
a vacuum being produced, the cushion is found to
be tightly pressed against the open end of the tube.
On gently releasing this, the mercury rushes up
with violent rapidity, completely filling the whole
space (see fig. 11); showing that neither oxygen
nor hydrogen remains in excess. The two gases
combine to form a trifling, in fact invisible, portion
of water. Such is the main principle upon which
the estimation of oxygen in atmospheric air de-

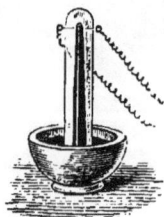

Fig. 11.
After the Explosion.

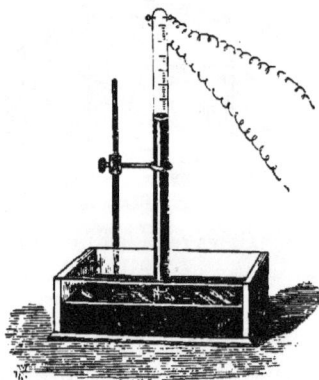

Fig. 12.

pends. In practice long tubes (fig. 12) are taken
very accurately marked into an immense number
(about 700) of small and equal divisions by an
engraving instrument attached to a dividing engine.
A specimen of air is introduced (2 or 3 cubic
inches) after having attained the temperature of

the room ; the level of the mercury in the tube is carefully measured by means of a telescope fixed perfectly horizontal on a vertical stand ; at the same time the temperature and height of the barometer are noted. Hydrogen is then let up into the tube to mix with the air in the proportion of about half its volume ; the bulk of the mixture is again measured, the barometer and thermometer again observed. Next a spark is passed through the gas, whereby a certain amount of contraction takes place. Another set of readings as before are made, and the loss in volume of the original quantity of air is the volume of oxygen which has disappeared in company with twice as much hydrogen to form water. The barometric pressure and the temperature are necessary observations, because changes indicated by these instruments affect the volumes of gases to a great extent. The error made by an experimenter skilled in gas analysis never exceeds $\frac{1}{10000}$, and most frequently is not more than $\frac{1}{100000}$

Lewy analysed many hundreds of samples of air collected in Paris, at Havre, on the Atlantic, in New Granada, and on many mountain-tops. Regnault in 1852 published his 'Researches on the Composition of the Atmosphere' in the 'Annales de Chimie.' His plan was to send out instructions to his friends and acquaintances, and others willing to help him in various parts of the world, to collect samples of air, to be sent to him for analysis. These were taken always at the same hour, towards noon, and on the first and fifteenth of each month during a whole year. At the same time the air of Paris was collected and submitted to analysis in the same way and with the same apparatus. The way in which the specimens of air were taken and preserved was the following :—Glass tubes were used, ending

Fig. 13.   Tube for a Sample of Air.

in two open points, these drawn-out points being very fragile. To avoid being broken in carriage the ends are covered by two little capsules, which are fixed on with wax or gum mastic, as in fig. 13. Each tube thus protected is placed in a cardboard case. To take a sample of air,

Fig. 14.   Transference of Air to the Tube.

soften the wax or mastic, detach the capsules, and connect one of the points with a pair of bellows by means of a piece of india-rubber tube ; then blow gently for two or three minutes (fig. 14).

The air of the tube is thus replaced by that of the bellows, which takes its supply from the surrounding atmosphere. It is necessary now to seal the tube hermetically. For that purpose withdraw the point from the india-rubber and heat it in the upper part of a spirit-lamp flame. When the

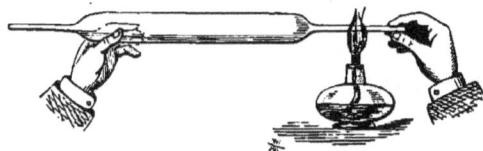

Fig. 15. Hermetically sealing a Tube.

glass is softened the point is drawn gently out so as to detach it from the tube, but without removal from the flame, and thus the tube is sealed on that side; the same operation is then performed at the other end. Specimens were taken in Berlin, Madrid, Geneva, Chamounix, on the Mediterranean, in the South Seas, and during the Arctic voyage of Captain James Ross[1] in search of Sir John Franklin.

The conclusions arrived at were:

1st. That the atmosphere shows sensible, although very small, variations in its composition, for the quantity of oxygen does not generally vary more than 20·9 to 21 volumes in 100 of air; but that in some cases, observed more frequently in

[1] Since Sir James Ross.

warm countries, the proportion of oxygen is as low as 20·3 volumes per 100 of air.

2nd. That the mean quantity of oxygen in 100 volumes of atmospheric air in Paris during the year 1848 was 20·96.

Taking into consideration the observations of Bunsen, Lewy, and Regnault together, we may consider the following fact established :—*Air from whatever altitude, and from whatever spot on the earth's surface, is very nearly of uniform composition.* This the following table from analyses of a later observer, Dr. Angus Smith, makes evident :

		Volumes of Oxygen in 100 of Air
Sea shore and open heath. Scotland	.	20·999
Tops of hills. Scotland . . .	.	20·980
In a suburb of Manchester. Wet weather.	.	20·980
The same	.	20·960
St. John's, Antigua	.	20·950

We see then the constancy of composition of the atmosphere notwithstanding that its constituent gases are not of the same weight, oxygen being a little heavier than nitrogen in the proportion of 16 to 14. This difference in weight may be well shown by the following arrangement: A jar af oxygen O, closed by a glass plate, is placed upon the table ; a jar of nitrogen N, open top and bottom, both openings being closed by glass plates, is placed over it (see fig. 16), so that the two gases may come together when the glass plates in contact are removed.   The nitrogen will float for a few seconds above the oxygen, and then if a lighted taper be quickly introduced through the neck of the upper jar it is extinguished in passing through the nitrogen, rekindling brilliantly on reaching the oxygen in the lower jar, so that there can be no doubt as to the difference in weight of the two gases. It is possible to account for the constancy of composition by assum-

Fig. 16.

ing, as Döbereiner and Prout did, that the oxygen and nitrogen in the air are chemically combined, but then we must, before making such an assumption, assign good reasons for it.   For the purpose of considering this matter let us examine the difference between substances which are chemically combined and mechanically mixed.

To this end take the red gas nitric peroxide, and see how different it is in all its properties to a mixture of its constituents. A large bottleful of this gas is seen to be of a reddish-brown colour, but in small quantities its tint may be described as orange-red. It is heavier than air ; so much so, that it can be poured out of the vessel containing it into another. When a lighted wax taper is plunged into it, it burns with a dull red flame, while a wooden spill is extinguished altogether ; yet there is in a bottleful of this gas a much larger proportion of oxygen and a smaller proportion of nitrogen than exists in an equal measure of atmospheric air ; for while this red gas contains 2 volumes of nitrogen and 4 volumes of oxygen, air, as already explained, contains 4 volumes of nitrogen, and only 1 volume of oxygen, and yet in atmospheric air the presence of oxygen is more evident than it is here from its behaviour towards combustible matter.

Furthermore, weight for weight the oxygen present in the brown gas is much greater, for whereas, 100 parts by weight of air contain 23 parts of oxygen, 100 parts of nitric peroxide contain more than $69\frac{1}{2}$ parts of the same gas.

Take as another comparison a mixture of 4 volumes of oxygen with 2 volumes of nitrogen ; it is colourless, and on introducing a lighted taper the combustion is not feeble, but most brilliant, and so likewise with a half-extinguished wooden spill. In fact, the properties of this kind of air are such as we should assign to a gas containing so large a proportion of oxygen, supposing that we

had never seen it ; things burn in it rather more
than three times as rapidly and brilliantly, because it contains rather more than three times
as much oxygen as air. But the reason the same
conditions are absent when we take the red nitric
peroxide gas is, that *when two substances chemically combine, the distinctive properties of each
generally disappear, and some new properties are always developed.* This is why chemical research
constantly discloses new substances with new
properties, and why in nature we have an almost
infinite variety of different kinds of matter, that is
to say, substances with different properties. Another
illustration of the difference between combined and
mixed materials is thus again afforded : two bottles
standing side by side may each contain a colourless
gaseous substance, composed of equal proportions
by bulk of oxygen and nitrogen. The mixture
behaves in no way different to one's expectations
when the stopper is removed, and on plunging in a
lighted taper it burns with brilliancy and indicates
the presence of oxygen in a large proportion. Now
another gas, nitric oxide, behaves in a very remarkable manner towards atmospheric air, so that when
the stopper is removed from a bottleful it actually
becomes red-brown in colour—in fact, is converted
into nitric peroxide—and on plunging in a lighted
taper it is extinguished. Yet another illustration :

D

one bottle containing a greenish-grey powder, and another a brilliant colourless, mobile, beautiful liquid, are each filled with the same materials. In each bottle there are 64 grammes of sulphur and 12 grammes of carbon ; in the one only intimately mixed, in the other they are chemically combined to form the liquid carbon disulphide.

It is now evident that when two substances are mixed the mixture partakes of the properties of each of the constituents, and the properties of the mixture vary as the proportions of the constituents vary. But when two substances are combined, the properties peculiar to each constituent disappear, and a set of entirely new properties are developed. Again, substances combine only in fixed and definite proportions, so that given any chemical compound, no matter how it be made, the proportions by weight of its constituents are invariable and its properties are invariable in consequence. In this lies the distinction between a compound and a mere mixture, in which the constituents may be present in any proportion whatever. It has already been demonstrated how oxygen and hydrogen gases contract in volume on combination to form water : such contraction is a frequent occurrence during chemical combination. In fact, chemical combination is made evident to us by the manifestation of one or more of the following phenomena :

1st. By disengagement of heat, as in the burning of phosphorus.

2nd. By the contraction of the constituents to a smaller volume.

3rd. By the resulting compound possessing properties which differ strikingly from those of its constituents.

4th. By the constituents of a compound occurring in fixed and invariable proportions. These are identical with, or bear a simple relation to, certain proportional weights which are found to regulate the combination of simple substances.

That air has none of the properties of a compound may be easily shown. Four pint bottles of nitrogen, and one pint of oxygen are simply mixed in a tall jar standing over water. No contraction in volume or evolution of heat takes place, and the product acts in every way as air. In short, the following are the arguments that air is not a compound but only a mixture :

1st. By the physical and chemical properties being the exact mean of those of its constituents ; so that in the easiest manner possible, knowing as we do the properties of the substances of which it is composed, we can by calculation, readily foretell he properties of air.

2nd. A much feebler chemical action is suffi-

D 2

cient to separate its constituents than is effective with most chemically combined substances.

3rd. Solution in water, which is not a chemical process, is capable of separating to a large extent the oxygen of the air from the nitrogen, so that air which has been dissolved in water is found to be richer in oxygen than undissolved air. Humboldt and Gay-Lussac found in the air

	In 100 parts of Air
Of distilled water . . .	32 parts of oxygen.
Of river water (Seine) . .	31·9 ,, ,,
Of rain water . . .	31·0 ,, ,,

But the experiments of Boussingault are very decisive and of great interest. It was observed by Bischoff that the air entangled in snow contained 10 to 11 per cent. only of oxygen. Objecting to these numbers, on account of the way in which the analysis had been conducted, Boussingault repeated them, and found that when snow was melted the air remaining undissolved by the snow-water had this composition, but the air contained in solution in the water had 32 per cent. of oxygen. But the neatness of the work was completed when he showed *that the composition of the air dissolved in the water and that of the air undissolved gave figures which represented the composition of the atmospheric air as determined on the same day.*

4th. By a process of filtration through thin india-rubber, it is possible to partially separate the oxygen from the nitrogen in atmospheric air.[1] (See fig. 17.)

5th. Air may be easily made by mixing oxygen and nitrogen in the suitable proportions, when it is found that no evidence of chemical or physical change takes place, such as accompanies the formation of a compound substance.

A problem well worth solving is the separation of oxygen from atmospheric air, so as to make it available for many purposes, such as, for instance, the production of very high temperatures. This has at last been successfully accomplished by M. Tessié du Motay. The

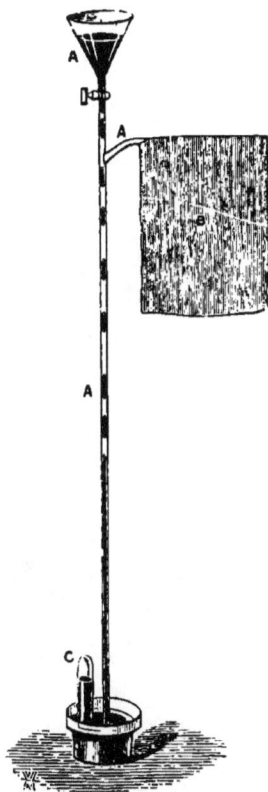

Fig. 17.

[1] The process may be explained by the aid of the diagram fig. 17. B is a flat bag made of Macintosh silk, containing a piece of carpet to prevent the two sides coming in close contact. It is attached to the Sprengel pump A A A, which, by reason of a fall of mercury from the funnel down the vertical tube, drives the air out and delivers it at C. Each drop of mercury acts as a piston.

process may be demonstrated in the following
manner (see fig. 18): a copper tube *t* is filled
with a mixture of manganate of soda and oxide
of copper, and placed in a furnace consisting of
a row of gas-burners *b*. One end of the tube is
connected with a T-shaped tube, so that by one
limb *a*, a current of air, or by the other a jet of

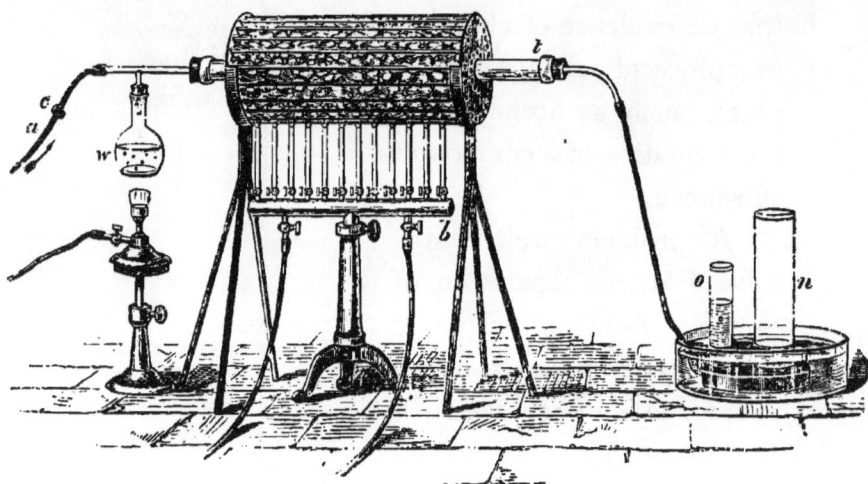

Fig. 18.   Apparatus for Extracting Oxygen from Air.

steam from the flask *w*, may be passed into the
copper tube.

On heating the manganate of soda to dull redness, and passing a current of steam through the

After a good vacuum is attained, the gas delivered at c has more
the properties of oxygen than atmospheric air; in fact, the percentage of this gas rises from 23 to 47.   This being the earliest and
simplest form of Sprengel's pump, shows its mode of action best.

tube, oxygen is liberated and collected at *o*, while within the apparatus caustic soda and oxide of manganese are produced. When the current of steam is shut off and air is turned on, the oxygen of the air is abstracted by the contents of the tube, while the nitrogen passes out at the extremity, and is collected in the cylinder *n*.

The formation and the instability of manganate of soda may be readily shown. In a silver dish supported over a gas burner a little caustic soda is fused. Add to it a little black oxide of manganese, and the formation of a beautiful rich green colour gives evidence of the production of manganate of soda. This green substance now dissolved in water, if added to a jar of alkaline liquid, remains green, while one or two drops in a gallon of dilute acid give rise to the splendid red colour like that of Condy's disinfectant, which is the permanganate of potash. Why the air is of such constant composition although its constituents are not of the same specific gravity, and are not combined, will be shown in a future chapter.

CHAPTER II.

AT the University of Edinburgh it is usual for graduates in medicine to contribute a dissertation on some scientific subject containing original observations and research. In 1754 the inaugural thesis of Dr. Black contained matter of such importance, treated in a manner so original, that it has raised the inexact observations of the alchemists and older chemists to the dignity of a true science. He showed that quicklime when exposed to the air becomes mild or carbonated, and at the same time increases in weight ; that this increase of weight depends on the absorption from the atmosphere of a kind of air or gas (carbonic acid), which gas could be again expelled or recovered by heating the lime to redness ; and that the increase of weight of the lime was identical with the weight of the gas absorbed. This is the first instance of the employment of the balance in the investigation of chemical changes, the first beginning of quantitative research, and as chemistry is a science founded entirely on observations of the weights and volumes of matter participating in change of properties, it may emphatically be said that Joseph Black was the Father of Chemical Science, and that Edinburgh University was its birthplace. The classic research of Lavoisier, already mentioned, by which he proved the composition of air, is but an application of Black's mode of thought and experimenting to other substances. There can be little doubt now, since the correspondence between these two illustrious men [1]—dating 1789-90—has come to light, that Lavoisier was accustomed to look up to Black as his master, and regard his discoveries as leading to great reforms in chemical knowledge.

Black further observed that the gas absorbed by lime could be expelled by the addition of acids.

[1] Mentioned by Prof. Andrews at the Edinburgh meeting of the British Association. See Report for 1871,

He compared this air with that evolved during fermentation, and gave it the name of fixed air. After explaining the difference between caustic and mild alkalies as being due to the presence or absence of this fixed air, he next showed that it differed from atmospheric air in its chemical character and its relations to respiration and combustion; furthermore, that it had the power of combining with alkalies, and to a feeble extent neutralising them after the manner of acids, and finally that it was, in fact, the product of respiration and combustion.

When lime is burnt in a kiln, or more correctly speaking, very strongly heated, the fixed air (or as it is now called, carbonic anhydride, carbon dioxide, or carbonic acid) is expelled, and the lime becomes caustic.

There is this great difference between mild or carbonated lime and caustic or quicklime : the latter is soluble to some extent in water, and possesses the property of turning red litmus blue and yellow turmeric paper brown, while the former, better known as chalk, is insoluble and shows no such effects. This difference affords a ready test for carbonic acid. The production of carbonic acid (as I shall always call it) by fermentation is easily exhibited in the apparatus shown in fig. 19, where the gas is evolved and collected over water during the

process of fermenting sugar with a little brewers'
yeast. That it is produced during combustion is

Fig. 19.   Collecting the Gas evolved during Fermentation.

easily demonstrated by taking some lime-water,
that is to say, a solution of caustic lime, and
shaking it in a bottle in which a taper has been
burnt : the clear liquid becomes turbid from the
formation of chalk or carbo-
nated lime. That it exists
in the air is easily shown by
placing a dish of lime-water
in a room, when it speedily
becomes coated with an insoluble layer of carbonate of
lime in the same way. That
it is produced during respiration is evident by the arrangement of bottles shown
in fig. 20 : they both contain
lime-water, and the tubes

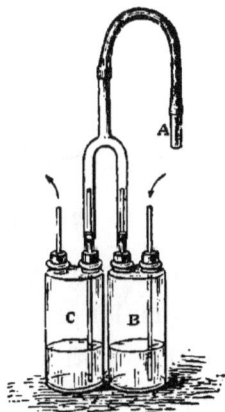

Fig. 20.   Apparatus
for showing Carbonic
Acid in the Breath.

connecting them are so arranged that by placing
the flexible one A in the mouth, and taking in

breath, air is drawn through the lime-water in the one vessel B, before entering the lungs, which causes little or no formation of chalk or carbonate of lime. By expelling the air from the lungs through the other vessel C a very considerable quantity of chalk is formed, as may be seen by the increased milkiness of the liquid.

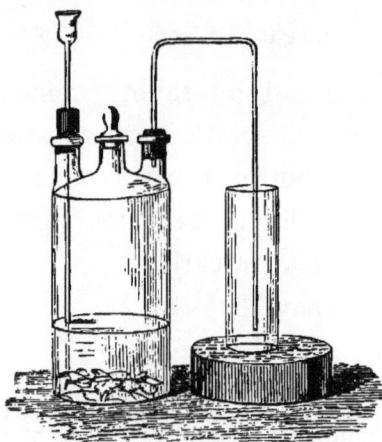

Fig. 21.   The Preparation of Carbonic Acid.

Having thus considered the various sources of this gas, let us examine its usual method of preparation and chief properties. It is usually made from marble by the action of hydrochloric acid, in a two-necked bottle, such as that in fig. 21. The gas is invisible, being colourless ; it is nearly odourless and tasteless. It is, however, more than half as heavy again as atmospheric air, and this may be easily made apparent by allowing it to flow

into a large jar: it will accumulate in the lower
part, and can then be cautiously ladled out, and,
by taking care, poured into a glass beaker, which
is carefully balanced ; the balance then descends
on that side into which the gas is poured, showing it to be heavier than the air which previously

Fig. 22. The Weight of Carbonic Acid shown.

filled the beaker (fig. 22). This may further be
demonstrated by allowing a light india-rubber
balloon full of air, or a soap bubble, to fall on to
a vessel filled with the gas (fig. 23) ; the soap
bubble will be seen to rebound from an invisible
surface of air, and then remain tranquilly floating.
This gas more resembles nitrogen than oxygen in

its non-support of combustion and power of extinguishing a flame.   Thus by ladling out some of

Fig. 23.   Soap Bubble floating on Carbonic Acid.

the gas we may, by taking advantage of its weight,
pour it some distance through the air on to a
candle and extinguish it.   The weight, too, and
non-support of combustion may be shown by
making a tree of lighted candles stand in a jar, to
the bottom of which a supply of carbonic acid is
allowed to flow by means of a tube ; as the gas
rises the candles are put out one by one.   Then
again a torch of blazing tow is easily quenched by
immersion in the gas (fig. 24).   Extinguishing
power consists in retarding the access of oxygen
or air and cooling the burning substance.   It is
not necessary to entirely prevent the contact of
the combustible with oxygen in order to extinguish
it ; mere dilution of the oxygen in atmospheric air
will suffice.   On account of the extinguishing power

of carbonic acid a candle cannot continue to burn in a portion of air until it has exhausted the oxygen, but only until the combustion has produced sufficient carbonic acid to extinguish the flame. This may be shown by enclosing a candle in a bell-jar standing on a plate containing water so as to ex-

Fig. 24.   Extinguishing Flame with Carbonic Acid.

clude air. On the plate and within the jar is placed a piece of phosphorus, and in the mouth of the vessel is a piece of hot bent wire. Now the candle burns till the product of its combustion causes it to flicker and go out. On turning round the jar so as to bring the hot wire and phosphorus in contact,

the latter will take fire and burn till it has consumed the remaining oxygen.

The other minute constituents of the atmosphere, besides carbonic acid, are water, ammonia,
and ozone. That water is present in the air may
be very easily shown by the effect of an ice-cold
vessel. A deposit of dew takes place, and if the
action is long continued a considerable quantity of
water may be collected. If the temperature is
lower than the freezing-point the deposit occurs
in the solid form as hoar frost, as is easily seen
when a mixture of ice and salt, having a temperature of $-16°$ C. or $0°$ F., is put in a blackened
metallic vessel. Now the source of this moisture
in the air is evaporation from the ocean, and the
extent to which it varies is due to temperature.
Water, like all other liquids, has the property of
giving off a portion of vapour at temperatures
below the boiling point, and as the temperature
rises the increase of vapour generated gradually
gets greater until the whole of the liquid is converted into vapour. When water exists in the air
as true vapour, it is, of course, invisible; when,
however, it is chilled, we get a portion deposited as
fog or snow. The extent to which water generally
exists in the atmosphere is 1·5 volumes of aqueous
vapour in 100 volumes of air.

This makes it a matter of great moment to

chemists, in the operations of analysis and other investigations, either to know precisely how much water vapour the air at the time of experimenting contains, or to have the means of getting air perfectly free from moisture. This latter end is arrived at by keeping oil of vitriol constantly exposed to the air. A piece of apparatus in constant use in chemical laboratories is a desiccator, that is, a bell-jar or other chamber, containing a vessel of oil of vitriol, over which is placed a floor of perforated zinc for substances which have to be kept dry. That atmospheric moisture is not entirely due to the existing oceans, lakes, and rivers, and the evaporation from their surface, is proved by the fact that a bell-glass held over a burning candle or gas jet is soon covered with condensed moisture, showing that it also is a product of combustion.

The presence of water vapour in the air is shown constantly by its effect on certain solid substances which have the property of combining with water, and in so doing becoming liquid. Chloride of calcium is one of these ; on account of this property it is said to be deliquescent. A compound of cobalt and iodine—called iodide of cobalt— when perfectly deprived of water at a temperature of 140° C., or 284° F., takes the appearance of masses of plumbago or black lead, being black and lustrous.

E

Exposure to the air on a plate soon resolves small
fragments of it into a rich green liquid, and this
green liquid after a time increases in volume and
becomes red, so that we have very pretty and
curious phenomena accompanying deliquescence in
this particular instance. A variation of the experiment may be made. A small quantity of the red
solution is smeared over the bottom of two white
porcelain dishes, in very small quantity ; so little,

Fig. 25.   The Dehydration of Iodide of Cobalt.

indeed, is necessary that the pink colour is scarcely
visible even on close inspection.   Heat the one dish
over a gas burner, and at first a yellow tinge appears,
but soon the whole of the bottom of the dish is
covered with a rich, moss-like green tint ; and now
in the centre of this a black spot appears, which
spreads gradually over the whole of the basin till it
has the appearance of being black-leaded (fig. 25).

By dipping the outside of the basin into water it is
cooled down, and while it is cooling the other dish
may be heated, and thus the reverse operations go
on side by side.   By heating, the colours appear; by
cooling, they disappear; and on close inspection it
will be seen that their disappearance has resulted
in the thin film of black shining substance being
transformed into a pink liquid.   The large, cool
surface of the dish readily affords a means of condensing atmospheric moisture, and the peculiar
properties of the cobalt iodide make this condensation visible.

This is the most remarkable deliquescent substance at present known.   When anhydrous, that
is, entirely deprived of water, it is black.   It
combines with a certain definite proportion of
water to form a green compound, and with a still
larger quantity to produce a red substance; hence
the curious change of colours when hydrated, or combined with water to a greater or less extent.   The
little instrument commonly sold in the opticians'
shops now, and called a chameleon barometer,
simply indicates the amount of moisture in the air
and in its action depends on the fact that paper
steeped in chloride of cobalt and chloride of calcium solutions has the property of becoming blue
in a dry atmosphere, and red when moisture is
present.

E 2

Without dwelling longer on the subject of atmospheric moisture, let us pass to the third minute constituent of the air, ammonia. This is a very light, colourless gas of a very pungent odour. Its specific gravity or relative weight, when compared with hydrogen, the lightest of all gases, is 8·5, or a little more than half that of atmospheric air. Its solubility in water is so great that at ordinary temperatures 1 volume of water absorbs 700 times its bulk of ammonia.

Notwithstanding the pungency of this gas and its extreme solubility, the fact of its being present in the air without our perceiving it is accounted for by its extremely minute proportion. This amounts to no more than $3\frac{1}{2}$ volumes in 10 million volumes of air. In other words, if we take 10 million gallons of air, the ammonia gas present would not be more than $3\frac{1}{2}$ gallons. To detect and accurately estimate the ammonia when in so small a proportion requires the use of very delicate chemical tests, very great care, and great patience in the performance of the experiments. From the numbers obtained by Fresenius, who is perhaps the highest authority we have on analytical matters, we learn that the ammonia present in the air varies by night and day, the amount being larger at night by about one-third. Mr. Horace Brown in 1870 published the latest experiments on the determina-

tion of the amount of ammonia in the air, and by
making use of the valuable process known as
Nessler's test, he obtained numbers which there
is every reason to believe are accurate—firstly,
because they are not widely different from those of
Fresenius, and, in fact, are closer to them than those
of any other experimenter; and secondly, the
numbers are just a little higher—a variation one
would be led to look for, as the method pursued by
Fresenius would be slightly in error, probably
from loss of ammonia. · Here, in the form of a table,
are the numbers given :—

Table showing the Ammonia in 10,000,000 *parts of Air.*

HORACE BROWN, 1870.

1869					By weight	By volume
Dec. 6	.	.	.	.	2·084 parts	3·542 parts
,, 8	.	.	.	.	2·15	3·65
,, 9	.	.	.	.	1·805	3·068
,, 11	.	.	.	.	1·812	3·18
,, 12	.	.	.	.	2·19	3·72

FRESENIUS.

By day	.	.	.98	1·666
By night	.	.	1·69	2·87
Mean	.	.	1·33	2·26

One remarkable property of ammonia is, its
power of combining with acids to form solid compounds, called salts. This is shown by the following illustration. Two large glass globes contain

each about a gallon of a gaseous substance of
very opposite properties; in the one is ammonia,
in the other hydrochloric acid. They are at first
closed by glass plates, but by bringing these in
contact and gently removing them the two gases
rush together (fig. 26) with great violence, combining instantly under great evolution of heat to

Fig. 26. The Combination of Ammonia with Hydrochloric Acid.

produce a white, solid, snow-like substance known
as sal-ammoniac, ammonium chloride, or ammonia
hydrochlorate, a substance in many respects resembling common salt.

By passing air through long tubes containing
broken glass moistened with weak liquid acid, the
ammonia may be abstracted, and so condensed in

the form of a solution. A very striking experiment
arranged to show the exceeding solubility of ammonia gas in water is the following.   The large glass
flask (fig. 27), which has its neck inverted below
water, coloured red with litmus, is full of ammonia,
but the mouth of the flask is closed from contact with
the water by means of quicksilver.   By raising the

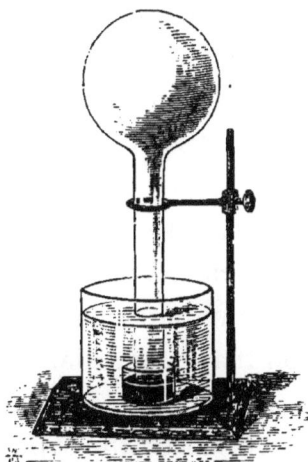

Fig. 27.   Experiment to show the Absorption of Ammonia by
Water.

neck of the flask out of the quicksilver into the water
the liquid dissolves the gas so rapidly, that it rushes
upwards with an alarming rapidity, the violence
being almost sufficient to break the vessel ; at the
same time it will be seen that the red litmus
solution becomes blue by the action of the
ammonia.

It cannot be doubted that the ammonia present in air is consumed by living plants ; indeed, it is this consumption that causes the difference in the amounts found by day and night. This difference was found by Fresenius to be in the ratio of 1 for the daytime and 1·7 for the night. The smaller quantity present during the day may be due to two causes :   1st. During the daytime more ammonia is absorbed in the nutrition of plants ; 2nd. The ammonia accumulating during the day and night together is dissolved and precipitated by the dew at sunrise.   .

The next and last gaseous substance to be considered as a constituent of the air is ozone. It is present in only very minute quantities, although in greater proportion than ammonia. Its nature and properties are most remarkable. The researches of many chemists have placed the supposition beyond doubt that it is a condensed form of oxygen with greatly exalted chemical activity. The energetic nature or oxidising power of ozone surpasses that of oxygen as much as the activity of oxygen excels that of air. The preparation of ozone on the lecture table in order that some of its properties may readily be demonstrated is thus effected (see fig. 28) :—The oxygen used is delivered from an iron cylinder of the compressed gas, but this is merely a matter of convenience and not of necessity.

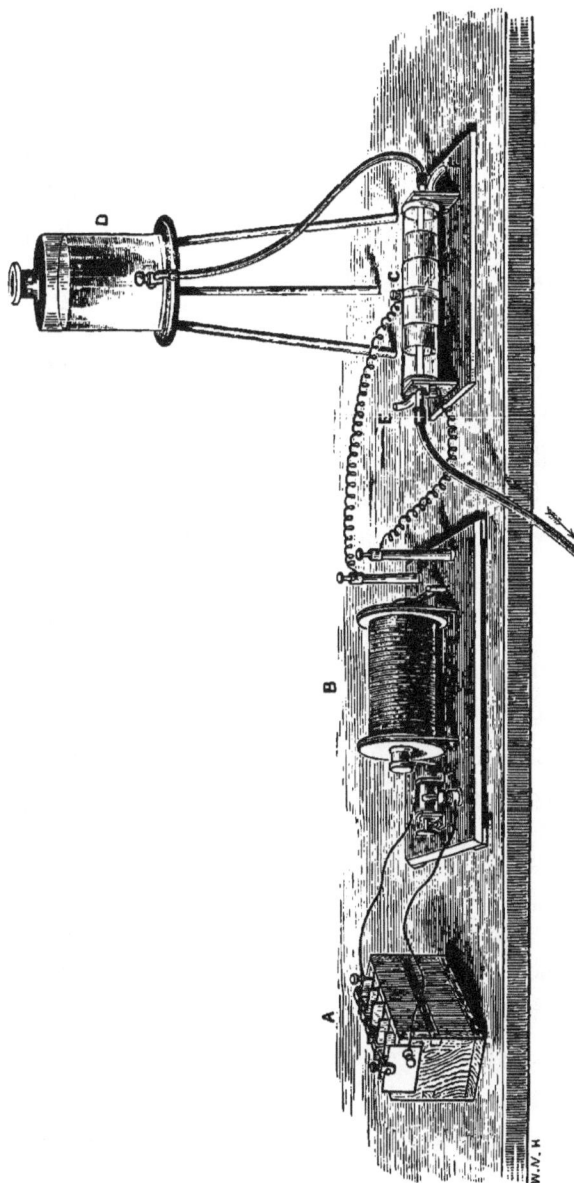

Fig. 28. The Preparation of Ozone.

A Galvanic battery to excite the coil. B Induction coil to produce an electric discharge. C Ozonising tube. D Reservoir of ice-cold water to cool the ozone tube. E and F Tubes for the passage of air or oxygen to be ozonised.

This gas when allowed to flow out is cooled by ice, afterwards dried with oil of vitriol, and passed into an apparatus known as an ozone generator.

The particular form of apparatus made by Messrs. Tisley and Spiller (fig. 29) is the one by which the best effects are produced. It consists of an iron tube turned very truly on the outside ; within it passes a current of water by means of the tubes C C. Outside this metal cylinder is one of glass, which is only a very slight degree larger, and by an arrangement of tubes D D, the cooled air from the iron bottle can be passed through the annular space between the glass and metal cylinders. Part of the outer surface of the glass cylinder is covered with tin foil G. Now the outer tin-foil coating and the inner metal cylinder are connected with the two fine wires, from a Ruhmkorff induction coil, at the points E and F. The coil furnishes a constant electrical discharge ; this discharge passes through the annular space before mentioned, and the electrisation of the current of oxygen or air endows it with new properties, which are very curious. It now possesses a very peculiar smell, a smell such as accompanies the sparks from an electrical machine, and which is often recognised when two stones are broken together with the emission of sparks. This odour is no doubt what is perceptible near a place which has been struck by lightning, and which is

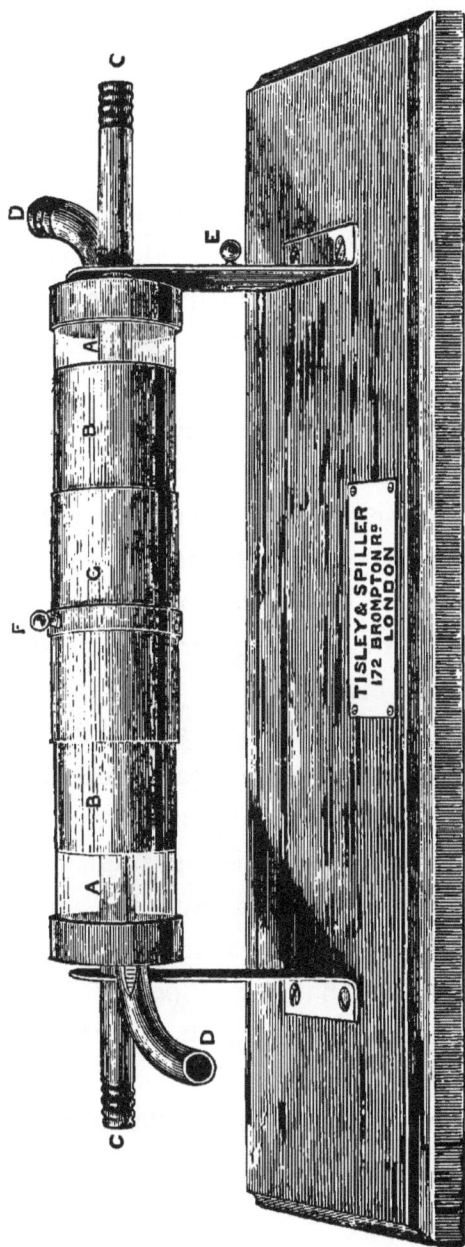

Fig. 29.   New Ozone Generator.

Woodcut from 'Nature,' by permission of Messrs. Macmillan and Co.

ignorantly described as a smell of sulphur. Besides this peculiarity, the gas has the property of attacking the mucous membrane of the nose and throat, giving rise to a painful soreness. It eats through any india-rubber tube with which it comes in contact for a few minutes, it has the power of bleaching indigo, it oxidises silver and mercury, and none of all these properties are possessed by common oxygen. It exercises its powers in such an energetic manner that the smallest possible quantity can readily be detected. There is one particular action which is peculiarly adapted for its ready discovery ; it sets iodine free from iodide of potassium. Now the merest trace of iodine in the free state has a remarkable power of striking an intensely blue tint when in contact with boiled starch ; so that by placing starch paste and iodide of potassium together on paper, we have a mixture which if exposed to ozone acquires a deep blue colour. The depth of the blue depends on the amount of iodine liberated, and the amount of iodine liberated corresponds to the quantity of ozone present, so that by exposing iodised starch-paper to a current of air in a dark place, if equal amounts of air are taken we can by the varying depths of tint compare the amount of ozone in one place with that in another, or the quantity present on one day with that on another occasion.

For some time it was by no means certain, after all said and done, that ozone was really present in the air. Nitric peroxide and nitric acid, also chlorine, could act on iodised starch-paper and cause the same blue colour to appear.

The matter was finally set at rest, however, both by Professor Andrews of Belfast and Professor Schoenbein of Basle in the same year, 1867, and by different methods. Andrews had previously shown that ozone, whatever be the method of its production, is quickly destroyed at the temperature of 237° C., or 458° F.

An apparatus was therefore fitted up by means of which a stream of atmospheric air could be heated to 260° C., or 500° F., in a large globular glass vessel. On leaving this vessel, the air was passed through a U-tube, the sides of which were moistened internally with water, while the tube itself was cooled by being immersed in a vessel of cold water. On passing air through this apparatus at the rate of three litres per minute, the test-paper of iodised starch was distinctly tinged in two or three minutes, provided no heat was applied to the glass globe. But when the temperature of the air, as it passed through the globe, was maintained at 260° C., not the slightest action occurred upon the test-paper, however long the current continued to pass. On the other hand, when small quantities of chlorine or nitric acid vapour, largely diluted with air, were drawn through the same apparatus, the test-paper was equally affected, whether the glass globe was heated or not. Hence the body in the atmosphere which decomposes iodide of potassium is identical with ozone.

The proof employed by Schoenbein was of a different nature. When white protoxide of thallium is exposed to the action of ozonised air, it is converted into brown peroxide. If we take a solution of sulphate of thallium to which a considerable amount of potash has been added so as to throw down the protoxide of thallium, papers dipped in this and exposed to the ozone instantly become brown; but it might be justly stated that if such papers be exposed to the air, a small quantity of sulphuretted hydrogen, a gas which escapes from decomposing organic matter, would also cause a similar effect. The difference between the browning in the two cases is this, that when due to peroxide of thallium, a paper steeped in tincture of guaiacum if pressed against it becomes blue where brought into contact with the brown thallium stain. The blue is the result of oxidation, caused by the peroxide of thallium; and it is needless to say that any other kind of brown stain would fail to show such a reaction with guaiacum solution. This thallium test is not so sensitive as the starch and iodide of potassium. Schoenbein notices a curious fact, namely, that the air is very rich in ozone after heavy falls of snow; this he remarked especially on January 6, 1867, at Basle, when he detected ozone by means of the thallium test.

A further most valuable contribution to our knowledge of atmospheric ozone is afforded by the researches of M. Houzeau, who makes use of a new test for this substance. This consists of red litmus papers, half of which are dipped in a solution containing one per cent. of neutral iodide of potassium. Should this paper come in contact with ammonia gas it will entirely assume a blue tint; with chlorine or acids it will be unacted on, while with ozone only that part of the paper will turn blue which is impregnated with the iodide of potassium. This is occasioned by the oxidation of the potassium to potash, the alkali thus formed changing the colour of the litmus. Houzeau's most important conclusions are the following :—

1st. That country air contains an odorous oxidising substance, with the power of bleaching blue litmus without previously reddening it, of destroying bad smells, and of blueing iodised red litmus.

2nd. That this substance is ozone.

3rd. That the amount of ozone in the air at different times and places is variable, but this is at most $\frac{1}{700000}$ of its volume, or 1 volume of ozone in 700,000 of air.

4th. That ozone is found much more frequently in the country than in towns.

5th. That ozone is in greatest quantity in spring, less in summer, diminishes in autumn, and is least in winter.

6th. It is most frequently detected on rainy days and during great atmospheric disturbances.

7th. That atmospheric electricity is apparently the great generator of ozone.

It has already been stated that ozone is a condensed form of oxygen. This was originally shown by the experiments of Professors Andrews and Tait.

It has, indeed, been well ascertained that 3 volumes of oxygen become condensed by electrisation to 2 volumes of ozone, and by heat we can permanently expand it again into 3 volumes of common oxygen.[1] Just as we have carbon exhibiting different properties in the three forms of char-

---

[1] Dr. Odling's explanation of this will be comprehended by those only who are acquainted with the atomic theory. While the union of only two atoms or ultimate particles makes a molecule of oxygen, the molecule of ozone contains three such atoms. Since the researches of M. Soret and Sir Benjamin Brodie, this view has been generally received.

coal or lampblack, plumbago or black lead, and
diamond, so we have
oxygen in the two forms
of common oxygen and
ozone. Such modifica-
tions of a simple sub-
stance are called allo-
tropic conditions.

It is most probable
that during thunder-
storms the violent or
disruptive discharge of
electricity from the
clouds causes, besides
ozone, a production of
nitric acid, or at any
rate of nitric peroxide,
which, in contact with
water, becomes nitrous
and nitric acids. The
evidence we have is,
that rain-water contains
minute traces of these
substances, and that by
making a miniature

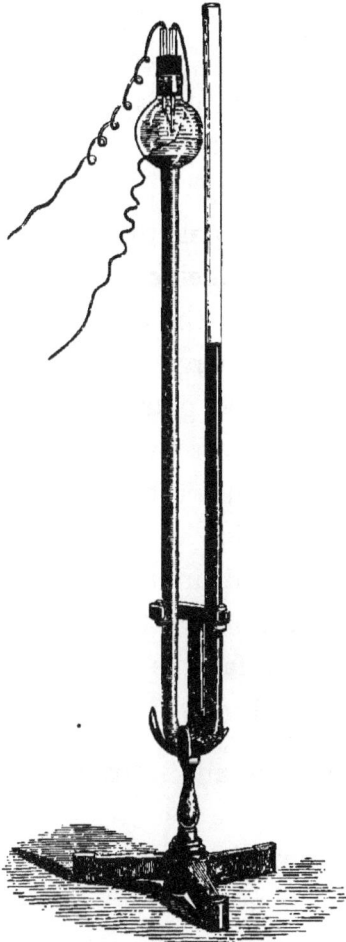

Fig. 30. Method of showing the
Effect of Electric Sparks on Air.

thunderstorm—as may be done by passing an
electric discharge through a glass bulb (fig. 30)—
the red fumes of nitric peroxide are rendered visible

F

Mr. J. M. Thomson, of King's College, has been engaged on experiments in this direction, and he finds that if a certain bulk of air be enclosed in such an apparatus as this, by passing electric sparks for a sufficient time all the oxygen of the air combines with enough of the nitrogen to produce the red nitric peroxide gas, the absorption of which by the water in the tubes causes a considerable contraction in volume of the original quantity of the air. When the discharge has gone on for twenty minutes within the glass bulb, the contraction amounts to that shown by the lower level of the liquid in the right-hand tube, fig. 30. The blue litmus, too, with which the tube is charged, though remaining unaltered in the right, becomes reddened in the left limb from the production of acid in the bulb.

So far the gaseous constituents of air may be divided into—

Oxygen and nitrogen   . Chief constituents.

Water   .   .   .   . ⎫
Ammonia .   .   .   . ⎪
Carbonic acids .   .   . ⎬ Constituents present in minute quantities.
Ozone   .   .   .   . ⎪
Nitrous and nitric acids   . ⎭

Besides these there are other gases present, occasionally, in more or less quantity, which may be called adventitious constituents, as hydrochloric acid from volcanic eruptions and chemical works ; sulphurous acid from the combustion of coal containing sulphur ; sulphuretted hydrogen from decaying organic matter ; and marsh-gas from the same source.

Returning to the consideration of carbonic acid and the amount present in the air, it has been observed that even in open places, many causes, amongst which is the state of the weather, give rise to very considerable variations in quantity, which are confined within very small limits.

To illustrate this variation are here tabulated some of the results obtained by Dr. Angus Smith, for a complete account of which reference must be made to his work on 'Air and Rain.'

	Volumes of Carbonic Acid in 10,000 Volumes of Air
On the mountains and moors of Scotland—mean of 57 analyses	3·36
In the streets of Glasgow—mean of 42 analyses .	5·02
London, N., N.E., and N.W. postal districts—mean of 12 analyses	4·445
London, S. and S.W. districts—mean of 30 analyses	4·394
London, E. and E.C.—mean of 12 analyses . .	4·745
London, W. and W.C.	4·115
Manchester streets, ordinary weather . . .	4·03
During fogs in Manchester	6·79

Here 3·36 volumes of carbonic acid in 10,000 parts of air may be taken as the normal amount. In towns the amount rises, and is greatest in those streets or parts which are the most populous. The state of the weather, again, influences the amount of carbonic acid ; during fogs it may be increased to 8 volumes per 10,000.

It was stated by Dalton that in his day the air of Manchester could not be distinguished by chemical analysis from that of Helvellyn, but Angus Smith has given us analyses which not only show the differences between town and country air—between the air of one town and that of another—but between that of various London streets, and even the variation in the air between the back and front doors of the house where his laboratory is situated in Manchester.

We have thus made evident the very great variability within very small limits, and the extraordinary constancy in composition of the air. All those who are susceptible to atmospheric influences prefer mountain air to that of the town, and in London prefer the air of the West End and the suburbs to that of the City.

From the mere evidence of one's sensations, it is impossible to say that the physiological effect which causes life to be more enjoyable in one place than another is due to the air we breathe ; but it is now placed beyond conjecture by these very exact chemical analyses that such is actually the case.   A vitiated atmosphere pervading the streets and hanging over cities is undoubtedly the chief cause of that longing for the country experienced by all dwellers in towns.

In the finest summer weather or in the depth of winter no one who takes the trouble to drive from Bayswater to Whitechapel could, I think, fail to notice the deterioration in the air as the more thickly populated parts are reached, and chemical analysis alone has shown us the cause of this. In foggy weather, when the air of a town is not carried away by a wind, but hangs over the house-tops like a cloak, the carbonic acid is not disposed of by diffusion and gentle currents of heated air with the same rapidity with which it is produced, and it therefore accumulates, even to a very unwholesome amount. At the same time the condensing aqueous vapour of which the fog consists carries down sooty particles and tarry matter from partially burnt coal, which greatly aggravates the other evil.

There can have been but few people in London last November (1873) who did not during the heavy fogs experience some difficulty in breathing, in getting a sufficient supply of oxygen for the lungs to carry on the vital processes in the usual manner. On one of the last days of its duration many people one met in the streets seemed to find breathing painful; they were apparently now and then gasping for breath, and the thought occurred

that it was by no means impossible by a conti-
nuance of such weather for a whole cityful of
people to be suffocated. The mortality at that
time amongst the sick was very great, and the
state of the air caused the death of many prize-fed
beasts sent to the Cattle Show at the Agricultural
Hall at Islington. It would be wrong to suppose
that the air was impoverished of oxygen to such
an extent as to cause asphyxia, but a serious
amount of inflammation, and, perhaps, a partial
obstruction of the air passages of the lungs, was
caused by the inhalation of soot, tarry matter, and
chilly particles of water.

It must not be inferred from the determinations
of carbonic acid here made so much of, that there
is something especially poisonous or dangerous in
minute quantities of this gas ; this is not so much
the case as that it is a measure of other accompanying impurities in the air ; 'not that we know certainly of any positive evil, which it can do of itself
in these small quantities, but because it almost
always comes in bad company.' It should also be
borne in mind that for every increase in carbonic
acid there is a decrease in oxygen, so that we have
a double effect, a subtraction of the life-giving
principle of the air, and an addition of a somewhat
noxious substitute. It will be seen further on, that
under the worst conditions out of doors, fog alone

excepted, people constantly within doors submit themselves to a worse atmosphere ; it is on this account that the estimation of carbonic acid is of great importance.

The quantity of carbonic acid is estimated by placing the air in a graduated tube over mercury (see fig. 12). A ball of potash at the end of wire absorbs the gas, forming solid carbonate of potash, and causing a diminution in the volume of the air. Accurate measurement before and after the experiment, and corrections for alterations in temperature and barometric pressure enable the operator to arrive at the true amount of the gas removed by the potash. All Dr. Angus Smith's numbers were obtained in this way. The method of Pettenkofer which I have practised is somewhat different.

Two graduated tubes, each of which is divided into 600 equal parts, the division being engraved on the glass, are two of the most essential pieces of apparatus. These instruments are called burettes, and are used for what is called the volumetric method of analysis. Chemists generally find it more convenient to use the French weights and measures, and in accordance with that system each 10 divisions on the glass would hold 1 cubic centimètre of water, which has the weight of 1 gramme ; therefore the whole buretteful contains 160 cubic centimètres or 60 grammes. The instruments are filled with two different solutions, one oxalic acid and the other baryta water. The strength of the liquids is so adjusted that a measure of the acid will exactly neutralise an equal measure of the alkali. This neutralisation is thus determined. Taking, for example, 30 cubic centimètres of this baryta water, one drop removed from the flask with a glass rod leaves a brown stain upon a disc of yellow turmeric paper ; if, now, 29 cubic centimètres of oxalic acid be run into the baryta water and well mixed up, the brown coloration is still produced, but it is fainter ; but on adding oxalic acid drop by drop the brown stain finally ceases to be perceptible.

The end of the process is now come, the causticity of the baryta water has been entirely destroyed by the oxalic acid, or in other words it has been neutralised. Now the oxalic acid is made of such a strength that each cubic centimètre will produce exactly the same effect towards neutralising the baryta as $\frac{1}{1000}$th of a gramme weight, or 1 milligramme of carbonic acid.

The next part of the process is to take an accurately measured flask, pump air into this from the desired locality with a pair of bellows, and then introduce from a burette 60 cubic centimètres of baryta water. An air-tight cap is then fitted on, the baryta water is swilled round the sides of the flask, and it is allowed to stand for at least two and not longer than twenty-four hours. The baryta water is then found to be quite milky, all the carbonic acid having gone to form carbonate of baryta. Pour out the turbid liquid, and allow it to settle, measure off half of the clear solution, and carefully proceed, as in the first operation, to add oxalic acid, and test with turmeric paper. If now 20 cubic centimètres of oxalic acid neutralise the liquid and only half the baryta water was experimented on, 40 cubic centimètres would have been the whole quantity required; this would indicate that 20 cubic centimètres had been neutralised by the carbonic acid. Such is an outline of the process. Its drawback is the bulkiness of the air-flasks, otherwise it is very convenient and very accurate.[1]

It has already been shown, that the outside air contains an amount of carbonic acid varying between 3 and 6 parts in 10,000 volumes, but it will be seen later on, in close places, such as crowded buildings, this rises to the extent of even 25 volumes in 10,000 of air. This, if the pollution of the air consisted of merely the addition of carbonic acid, would be a sufficiently alarming amount, but it is not; there is likewise the abstraction of oxygen to be taken into account, and the various exhalations from the lungs which ought to be regarded as poison.

[1] Fuller details are given further on.

It has been experimentally proved that *when the heat is not excessive the organic matter charging the air of crowded places rises in amount as the carbonic acid increases*, so that in the estimation of carbonic acid we have a measure of the foulness of the air or, as it may be termed, want of ventilation. Coming from the outside into a room in the condition we call 'close' or 'stuffy,' we enter an atmosphere which does not contain less than 6 volumes of carbonic acid in 10,000 of air. But the 'closeness' is detected generally by the nose, and is the effect of organic exhalations, rather than of carbonic acid ; nevertheless the two are hand-in-glove, so the carbonic acid, which can be measured with greater certainty and ease than the other pollution, tells the story for both.

A very easy method of finding the extent to which the air of dwelling-houses is contaminated has been devised by Dr. Angus Smith. This 'household' method of testing the air depends upon the fact that a certain definite quantity of carbonic acid is necessary to cause a visible precipitate in half an ounce of lime water. A practical illustration will be the best explanation possible. Taking five wide-mouthed bottles measuring respectively 8·05 ounces, 9·13, 10·57, 12·58, 15·6, or in cubic centimètres 228, 259, 299, 356, 443, if into the first or smallest there be put half an ounce

of perfectly clear lime water, the bottle being previously wiped quite clean and dry, and if after
shaking so as to mix the air and lime water, on examining it carefully, there be 'no precipitate,' pass
on to the next bottle a size larger ; if with the third
it is somewhat doubtful, but with the fourth bottle
the precipitate is well seen, let us take the exact
point to be between the third and fourth bottles.
Now on referring to a table previously constructed,
we find that this indicates between 5 and 6 volumes
of carbonic acid in 10,000 of air, say 5·5, *i.e.* $5\frac{1}{2}$—an
amount very likely to be present, and which indi-
cates a fairly ventilated place.

'*Let us keep our rooms so that the air gives no
precipitate when a* $10\frac{1}{2}$-*ounce bottleful is shaken with
half an ounce of clear lime water.*' We shall thus
secure good ventilation, the carbonic acid will be
less than 6 per 10,000. For sleeping apartments
such a rule is most valuable, for it is there we spend
about one-third of our lives, and during sleep we
have great need of all the oxygen we can get ; for,
according to Pettenkofer and Voit, the body is then
engaged in storing it up for future use.

The minimetric method of analysis of Dr. Angus
Smith resembles the already mentioned 'household
method' in its extreme simplicity. The principles
upon which it depends are, that a measurable
quantity of carbonic acid is required to produce

an appreciable turbidity in half an ounce of baryta water, and that the *exact appearance* of a certain precipitate of carbonate of baryta *can be retained in the memory with marvellous exactitude.* If we place half an ounce of baryta water in a bottle, having space for two ounces of air, and attach by means of an india-rubber cork and tubes a ball-syringe fitted with valves, and holding two ounces (see fig. 31), we have a means of pumping air in measures of two ounces at a time through the baryta until the particular precipitate most easily retained by the memory is produced. Such a precipitate is that yielded by shaking a 23-ounce bottle of air with half an ounce of clear baryta water, when the air contains 0·04 volume per cent., or 4 volumes of carbonic acid in 10,000 of air. The actual amount of carbonate of baryta formed is 0·00224 grammes. I have adopted the plan of making a solution which will always yield this precipitate. Dissolve 7·4536 grains or 0·4821 gramme of pure dry carbonate of soda—heated to redness in a crucible— in 5 pints of water; ¼ ounce carefully measured with a pipette will with ¼ ounce of baryta water solution produce the standard precipitate.[1]

When the ball-syringe is squeezed the con-

---

[1] Or the same quantity may be dissolved in 800 cubic centimètres of water, and only two cubic centimètres taken instead of ¼ ounce.

tained air is discharged by the tube with the free
end A ; on relaxing the pressure air enters from the
bottle by the india-rubber junction tube C ; at the
same time the external
air, passing down the
straight glass tube D
bubbles through the
baryta water.

In making an ana-
lysis first attach the sy-

Fig. 31.

ringe and cork to the empty bottle, then give two or
three strokes to get the air changed for that of the
place to be tested. Put in the baryta water and
shake it well; this counts for one ballful. Close the
entrance-tube with the forefinger, squeeze the ball,
remove the finger, and relax the pressure. When the
ball is fully expanded shake up again.   This is the
second ballful.   Repeat the operation as often as
is necessary until the liquid is of such a turbidity
as that produced by the standard precipitate.
Count the number of ballfuls of air, and refer to
one of the tables here given.

Number of ballfuls of air	Volumes of Carbonic Acid in 10,000 of Air			
	With a 2-oz. ball	1-oz. ball	½-oz. ball	¼-oz. ball. [1]
1	—	—	—	—
2	—	—	—	—
3	—	—	—	—
4	—	—	—	—
5	8·8	—	—	—
6	7·4	—	—	—
7	6·3	12·6	—	—
8	5·5	11·0	—	—
9	4·9	9·8	—	—
10	4·4	8·8	17·6	—
11	4·0	8·0	16·0	—
12	3·7	7·4	14·8	—
13	3·4	6·8	13·6	—
14	3·2	6·4	12·8	—
15	—	5·8	11·6	23·2
16	—	—	11·0	22·0
17	—	—	10·4	20·8
18	—	—	9·8	19·6
19	—	—	9·3	18·6
20	—	—	8·8	17·6
21	—	—	8·4	16·8
22	—	—	8·0	16·0
23	—	—	7·7	15·5
24	—	—	7·4	14·8

The accuracy of the test diminishes rapidly when a 2-ounce ball is used with air containing more than 7 volumes of carbonic acid per 10,000. A 1-ounce ball may be used up to 12·6 volumes per 10,000. It is an immense convenience to have the half-ounce measure for the baryta water marked on the glass with a diamond, and to engrave the suitable reference table upon the bottle itself. It

[1] The utility of a ball of this size is doubtful.

is also worthy of remark that a 1·ounce ball may
be used with a 2-ounce bottle, but not *vice versâ* ;
and under such conditions when the baryta water
is first shaken up instead of 1 this must count as
2 ballfuls. Again, it is a great advantage to have
a label written in lead pencil, of such a shade that
the standard precipitate just prevents its being
seen through the bottle, gummed on to one side.
If the words ' Not Done ' or ' Unfinished ' be taken,
as long as they are legible the operation is un-
completed. Without the standard solution of
carbonate of soda there is some inconvenience
attending the *exact* determination of carbonic
acid, though there is none in comparing the air
of a room with the air outside the house, and say-
ing that it contains twice, thrice, or four times as
much. Again, we may make an experiment on
the external air, and assume that it contains the
smallest probable quantity—for instance, of a
London street, say 4 per 10,000 volumes—test the
air of a room, and then by comparing it with that
of the street it is easy to fix the minimum of
carbonic acid present. For medical officers of
health and others engaged in sanitary inspec-
tions who cannot be expected to know anything
of analytical operations, especially those of such
a delicate nature as gas analysis, an apparatus
of this kind is invaluable in enabling them to

express an opinion as to the wholesomeness or
otherwise of an apartment, schoolroom, or public
building.

So great is its simplicity and accuracy that a
man who positively is ignorant of the operation he
is performing may yet estimate the carbonic acid
in the air of a room to the accuracy of 1 volume
in 10,000 if the quantity be large, and the error is
only half as great with small quantities. The
error, too, is on the safe side, the estimated amount
being rather under than over the actual quantity
with a cautious operator. When the air of a place
feels 'close' on first entering, a 1-ounce ball and
bottle is most convenient; if very 'close,' such as
the air of a crowded public building, the better size
is $\frac{1}{2}$ an ounce, the same quantity of baryta water
being taken in every case. Considerable experience
has shown that the silk valves used in the T-tubes
often cease to act properly, and the following
contrivance is found to give what is for a time a
serviceable and more easily constructed apparatus.
The ivory nozzle of a ball-syringe has about a foot
of black india-rubber fitted to it; the best size is
under $\frac{1}{4}$ of an inch in diameter. This has a longi-
tudinal slit made in it with a sharp penknife of
the length of an inch, and is attached to the bent
tube C (fig. 31) of the bottle, and on to the tube
D is fitted about $\frac{3}{4}$ of an inch of the same india-

rubber. By bending this piece of india-rubber with the finger all exit from the bottle is closed, but when the ball is squeezed air escapes by the slit in the connecting tube, but cannot re-enter until the finger is removed from D. This simpler form of apparatus has been found to be most efficient.

Considering the carbonic acid exhaled by the animal world, poured into the air by furnace fires, and slowly evolved during the decay of organic matter; remembering also that this process goes on without ceasing, is it possible in course of time that the oxygen now at our disposal will be so used up and reduced in quantity as to be no longer capable of supporting life? Dumas and Boussingault have made a calculation to answer this question, and they consider that under the worst possible conditions scarcely less than 800,000 years would have to elapse before the mass of the atmosphere would be deprived of its oxygen by animals living on the earth's surface, so vast is its extent. But the immense number of experiments of Angus Smith show numbers which exactly tally with those of Regnault, made twenty years previously: hence there is not the slightest dimunition in oxygen in so long a time, though the average error in Regnault's and Angus Smith's work is not greater than $\frac{1}{100000}$, so that in eight years we should be

G

able to detect some alteration in the numbers. It
is evident, then, that there is not merely an absorption of carbonic acid by rocks or the waters of the
ocean or rivers ; there must actually be a restitution
of the oxygen consumed in the production of carbonic acid. As Priestley originally discovered and
made evident, plant life on the earth's surface is
the cause of this. Just as a certain quantity of
oxygen is necessary for the maintenance of the
animal kingdom, so also an amount of carbonic
acid is required as food for plants. The respiration
of plants is effected through the leaves, which are
filled with stomata, apertures which may be likened
to small mouths ; by these they inhale carbonic
acid and exhale oxygen. The green colouring
matter of foliage or chlorophyll performs a similar
function, though the chemical process is different,
to that of the red blood corpuscles, for while the
red blood corpuscles are carriers of oxygen from
the air to the internal organisation of animals, the
chlorophyll acts as a supplier of carbon to the
tissues of plants ; the oxygen, not being wanted,
is returned to the air. The solar rays play an important part in this chemical action of plants, for it
is by the influence of sunlight that the nutritive
process goes on. Regarding the changes which
the oxygen undergoes, we have first its carbonisa-
tion and hydrogenation, by means of animal life,

and return to the air; then its decarbonisation by means of vegetable life, and restoration in its original state. Or if we consider the carbon in the same way, we have the consumption of vegetable products as food, the oxidation of animal tissues, and the oxidised carbon committed to the air, then again a reabsorption of the carbonic acid and its deoxidation in the form of vegetable products. So it follows that the air is a reservoir of oxygen for the use of animals, and a reservoir of carbonic acid for the use of plants.

In the round of operations by which the carbon of plants is derived from carbonic acid, water also plays as prominent a part. In carbonic acid and water there is contained far too much oxygen to admit of the building up directly of such substances as cellulose or woody fibre, sugar, and starch; by a process of condensation or coalition the carbon and hydrogen combine, while some of this oxygen is got rid of. The reverse operation takes place when we burn a piece of wood; the oxygen entering into its composition is insufficient by any re-arrangement with the carbon and hydrogen atoms to form again the substances from which it took its origin, but as the wood acquires oxygen from the air, its carbon and hydrogen atoms are separated, and it is again resolved into carbonic acid and water, and at the same time there is evolved an

amount of heat corresponding to the chemical
energy exerted by the sunlight when it promoted
the growth of the trees. Thus can we store and
concentrate the sun's rays so as to bring their
action to bear on any one particular place at any
given time. Such is the case where woods are
planted and grown for twenty years to be cut
down and thrown into a furnace. When primæval
vegetation in the form of coal gives the warm glow
of the fireside in winter it is the sunbeams of a
bygone age whose influence we feel.

APPENDIX TO CHAPTER II.

For those who are interested in the method of Pettenkofer for determining the amount of carbonic acid in the
air, here are given full directions, with some novel practical
details which simplify this process.

Pettenkofer's Method.—The flasks used for containing
the air are the ordinary long-necked bolt-heads ; they
should be able to hold 5 to 10 litres (1 to 1½ gallon).
Their capacity is determined by measuring in water at a
temperature of 16° C. (or 60° F.) out of a litre-flask.
When the water has risen up to the neck to a convenient
height, an additional 60 cubic centimètres are added. A
mark is then made on the glass at the level of the liquid
with a sharp file or cutting diamond, and the application
of a piece of hot glass or iron cuts off the superfluous
portion of the neck. The complete capacity of the

measured flask, supposing 7 litre-flasks full of water have been emptied, is 7,060 cubic centimètres ; its capacity for air, however, during an experiment is 7,000 cubic centimètres, because 60 cubic centimètres of baryta water are introduced.

Regarding the strength of the baryta water, this may be made by dissolving 7 grammes in 1 litre of water, if the baryta be the *crystalline hydrated* compound.[1]

The oxalic acid is made by dissolving 2·8636 grammes of pure crystallised oxalic acid in 1 litre of water. As 126 parts by weight of oxalic acid ($C_2H_2O_4$ ·2 Aq.) neutralise the same quantity of baryta or lime-water as 44 parts by weight of gaseous carbonic acid (carbonic anhydride), $\frac{126}{44} = 2\cdot8636$, the whole of this solution is of the same strength as 1 gramme of carbonic acid ; consequently, 1 cubic centimètre = 1 milligramme.

Take 60 cubic centimètres of the baryta water, measured with a pipette, and compare it in the following manner with the oxalic acid. Run into it 30 cubic centimètres of the acid solution, and stir with a glass rod ; if no permanent precipitate or turbidity is caused, the baryta should be diluted with an equal bulk of water. Supposing, however, that the liquid is no longer clear, take a drop out with a glass rod and let it touch a filter paper coloured with tincture of turmeric. As these test papers are bleached by exposure to light, it is best to store them away in an earthen jar in a dry place. Great care should be taken during the experiment to keep the air of the laboratory free from ammonia, even the quantity present in tobacco smoke is sufficient to vitiate the result. If a red or brown spot be seen, the baryta water is not

[1] That the baryta should contain no trace of potash or soda is of the greatest importance.

yet neutralised ; more oxalic acid is cautiously added, and
as in time the brown spots get fainter, increased caution
is necessary in the addition of the acid.   Finally, when
the liquid shows no action on the turmeric paper, the
baryta has been completely neutralised or saturated.   If
60 cubic centimètres of the acid have been used, then 1
cubic centimètre of acid $=$ 1 cubic centimètre of baryta.
It is rarely that this occurs, and it is not worth while
diluting or strengthening the solution so as to obtain very
exactly this relation of the two liquids.   If 90 cubic
centimètres of acid were required for 60 cubic centimètres
of baryta, then each cubic centimètre of the alkali would
be capable of neutralising $1\frac{1}{2}$ milligramme of carbonic
acid.   There is, therefore, as much alkali in 60 cubic
centimètres as there should be in 90 cubic centimètres,
and the liquid should be made up to this bulk.

To make the analysis, take the flask in a clean and
dry state, let it remain in the air for a quarter of an hour
to gain its proper temperature, then with a pair of bellows
blow air into it ; note the temperature, measure in 60
cubic centimètres of baryta solution, cover quickly with
a capsule of india-rubber, or tin plate and india-rubber,
such as are made for pickle and preserve jars, swill the
liquid round the sides of the glass, and let it remain from
two to four hours.   Pour out the liquid into a little
beaker glass or flask, cover with a watch glass, allow the
carbonate of baryta to subside, take off with a pipette 30
cubic centimètres of the clear liquid, and cautiously add
oxalic acid from the burette until the alkalinity is
destroyed.

Suppose it had been found in the first experiment
that 63 cubic centimètres of acid were equal to 60 cubic
centimètres of baryta, and that in the second case

26·5 cubic centimètres had been used ; then as only half the 60 cubic centimètres of baryta solution had been experimented on, multiply by 2 :

26·5 cub. cent. × 2 = 53 cub. cent. of oxalic acid.

But 63 cubic centimètres of oxalic acid is the total quantity 60 cubic centimètres of baryta can neutralise, therefore $63^{cc} - 53^{cc} = 10^{cc}$ of oxalic acid, or 10 milligrammes of carbonic acid.

Now refer to the following table, which gives the volumes of 1 milligramme of carbonic acid at different temperatures.

If the temperature had been 17° C., then—

·5403 × 10 = 5·403 cub. cent.

The capacity of the flask is 7,000 cubic centimètres, so the following proportion

7 : 10 :: 5·403 : x
7)54·03
7·71

gives the volumes of carbonic acid present in 10,000 of air. When several flasks are in use it is better not to mark their actual capacity on the glass, but the divisor and multiplier only, as in the above case, $\frac{1}{7}^0$.

Volume of 1 milligramme of Carbonic Acid at various temperatures.

°C.	Cubic centimètres	°C.	Cubic centimètres
At 0	= 0·50863	At 4	= 0·51608
1	= 0·51049	5	= 0·51794
2	= 0·51235	6	= 0·51980
3	= 0·51451	7	= 0·52167

Volume of 1 *milligramme of Carbonic Acid at various temperatures.*

°C.			Cubic centimètres	°C.			Cubic centimètres
At 8	.	.	= 0·52353	At 23	.	.	= 0·55177
9	.	.	= 0·52539	24	.	.	= 0·55334
10	.	.	= 0·52726	25	.	.	= 0·55520
11	.	.	= 0·52912	26	.	.	= 0·55706
12	.	.	= 0·53098	27	.	.	= 0·55893
13	.	.	= 0·53314	28	.	.	= 0·56079
14	.	.	= 0·53471	29	.	.	= 0·56265
15	.	.	= 0·53657	30	.	.	= 0·56451
16	.	.	= 0·53843	31	.	.	= 0·566
17	.	.	= 0·54030	32	.	.	= 0·568
18	.	.	= 0·54216	33	.	.	= 0·570
19	.	.	= 0·54402	34	.	.	= 0·572
20	.	.	= 0·54589	35	.	.	= 0·574
21	.	.	= 0·54775	36	.	.	= 0·576
22	.	.	= 0·54961	37	.	.	= 0·578

CHAPTER III.

The Means whereby a Constancy in Composition is maintained in
the Air—Currents produced by Unequal Temperatures—The Expansion of Gases by Heat—The Law of Charles—Dalton's Observations on the Mixing of Gases notwithstanding their Different
Specific Gravities—Graham's Law : the Rate of Diffusion of Gases
is Inversely Proportional to the Square Root of their Specific
Gravities—Explanation of the Reason of this by the Kinetic
Theory of Gases—Effects of Differences of Temperature on Diffusion—Of Ventilation, its Principles and Practice—Its Neglect
in most Public Buildings—Evil Effects of Foul Air—The Air
Permeating the Soil, or 'Ground-Air'—Passage of the Ground-
Air into Dwelling-houses—Danger from Leaky Drainage or
Gas-Pipes.

CONSIDERING the number of different gases possessing different properties, and having different
relative weights or specific gravities, which being
simply mixed together form the atmosphere, and
considering that some of these gases, such as hydrochloric acid, carbonic acid, and sulphurous acid, are
evolved in excessive quantities in certain districts,
such as manufacturing towns, how is it that in a
place like Glasgow, for instance, the composition
of the air is not greatly altered? Why does it
not contain an excessively large quantity of all the

noxious gases named above, more especially as
these substances are all much heavier than pure
air? Why does not the heavy carbonic acid form
the lowest stratum on the earth's surface, the
oxygen next, and the nitrogen lie over these, while
above all rests the ammonia? How is it that
Babinet's calculation and Dalton's theory concerning the decrease of oxygen in the air at great
elevations have no foundation in fact?

By studying the effect of rise of temperature on
gases, one half or more of the problem will be
solved. Take a bell-jar, and hold beneath it a roll
of brown paper which is smouldering and giving off
abundance of smoke ; this smoke may be observed
to rise and fill the bell-jar, that is to say, the air
being heated rises and carries the smoke along
with it. Having filled the bell-jar, remove the
stopper, and there is immediately an upward rush
outwards of heated air, which, by the disappearance
of the smoke, indicates that the contents of the
glass have been dispersed in the surrounding atmosphere. Again, by means of a charcoal fire one
may easily obtain a supply of heated air, which
consists for the most part of carbonic acid, a gas
at ordinary temperatures heavier than air. If this be
allowed to rise into an air balloon, it will so inflate
it and render it buoyant that it will be carried to a
considerable height before the gas cools. The

effect of rise of temperature is to cause gases to expand and become lighter.

The proportion in which they acquire this property is curiously related to the temperature to which they are heated. Thus on the Centigrade thermometer between the temperature of melting ice and that of boiling water there are 100 equal divisions or degrees. At 273 degrees below the melting ice temperature, we reach a degree of cold which has never been observed or produced artificially; but there is reason for supposing that a further decrease in temperature would cease to affect a gas in the same way that it does under all conditions at present known. This point is called the absolute zero, and the following statement is known as Charles' law:

The volume of a gas is directly proportional to its absolute temperature or temperature reckoned from the absolute zero. Thus if a gas measures 1,000 cubic centimètres at 16° C., its volume at 25° C. would be found in the following manner: first calculate the absolute temperature corresponding to 16°; this is 273° + 16 or 289°; then that at 25° or 273° + 25° = 298°, after which the following proportion will establish the change in volume · which would take place by this rise of 9° C. :

$$289° : 298° :: 1000^{cc} : 1031^{cc}$$

In other words, the effect of heat on gases may be

said to cause an expansion of $\frac{1}{273}$ of their bulk for each degree on the Centigrade scale. The effect of the sun on the earth's surface by heating the lower stratum of air causes it to be expanded so that it rises ; at the same time cooler air moves onward to supply its place. When the heating action is exerted on the ocean, evaporation from its surface takes place ; the vapour rises producing a similar effect. The heated air rising, the cooler would rush in a lateral direction, and if with considerable force this would constitute a wind. Winds are nothing more or less than such phenomena, but their direction is determined in certain cases by other causes. If, for example, the air between the tropics, where the sun's rays have greatest power, rushes upwards with great velocity, it is only natural that under so great a disturbance the air at the Poles, which is the coldest, should push forward to take its place. In doing this, however, it changes its entire position on the earth's surface.

Now, as already mentioned, the atmosphere rotates with the earth upon its axis, and the velocity of this rotation is of course smallest at the Poles ; in fact, we might fix on a particular point where it amounts to nothing ; but at the Equator the *surface of the earth* moves with a velocity of 1,000 miles per hour from west to east. If the wind starts from the North Pole in a due south direction, having no motion towards E. or W., it will soon enter upon a region of the earth's surface which is travelling briskly, owing to rotation, in an easterly direction. The Polar current we are considering, will gradually acquire a portion of this easterly motion, but will always have less of it than the surface of the earth over which it is passing. Thus the air left behind by one portion of the earth's surface will be overtaken by another, producing the effect of a wind from the east, which, when combined with its original motion in a southerly direction, gives us a N. E. wind.

As the current advances towards those latitudes where the motion of the surface due to rotation is much faster, this effect will be intensified, and a strong steady N. E. wind will be perceived. In like manner a current starting from the South Pole will be, relatively to the earth's surface, left behind, and converted into a strong and steady S. E. wind.

These two currents blow with little variation throughout the year, and are known as the trade winds. They are balanced by return currents which flow back, at a different elevation, from the equator to the pole. These latter flow at first high above the level of the trade wind, but as they advance toward the pole, they descend at times upon the surface of the earth, and in our own country give rise to the destructive but healthful gales from the S. W.

Such are the means whereby the air is thoroughly mixed, though it is not entirely by such violent means that a circulation is effected.[1] It is seldom we experience a day without some wind, and even when the air is apparently quite calm, and not the slightest ripple on the water or the movement of a leaf betrays the motion of the air, still there are currents ever ascending. Around our bodies, for instance, by means of delicate instruments, it has been proved that the animal heat causes a constant upward streaming of the air. If the smoke of a cigarette be carefully expelled from the mouth, it may be deposited in the centre of a room in a very small space measuring less than a cubic foot. On leaving it to itself it will speedily be dispersed, and if the door be opened it may even be smelt half over a moderately sized house after an interval of a minute. This is in reality due to the movements of the air, which carries the smoke along with it, and not to the motion of the smoke by itself. By bearing this in mind, the extraordinary way in which the so-called still air circulates is readily comprehended.

[1] It is evident that the atmospheric tides must to some extent vary the direction and force of the wind, and if there be any truth in the popular notion that a new moon brings a change of weather, it is evident how the change may possibly be effected.

It has already been pointed out how the
reciprocal action of plants and animals preserves a
constant balance between the carbonic acid and
oxygen of the air. It is, necessary now to notice
the last cause operating so as to maintain a constancy in its composition. Dalton, who, as before
mentioned, supposed the air of different altitudes
to consist of different definite mixtures, completed
in 1803 an investigation on the tendency of elastic
fluids to diffuse through each other, and his results
established this remarkable fact, that a lighter
elastic fluid cannot rest upon a heavier, as is the

Fig. 32.   Interdiffusion of Hydrogen and Oxygen.

case with liquids ; but they are constantly active
in diffusing themselves through each other until a
final equilibrium is established ; and that without
regard to their specific gravities, except so far as
they accelerate or retard the effect according to circumstances. An effective illustration of this consists in taking two cylinders, one of hydrogen, the
other of oxygen. The specific gravities of hydrogen
and oxygen are 1 and 16 respectively, that is to
say, oxygen weighs 16 times heavier than an equal

bulk of hydrogen. Place the heavier oxygen down, and on the top of this the lighter hydrogen, and now remove the glass plates dividing them (fig. 32). In about half a minute one is able to show that these two gases are no longer separated, for by applying a light each cylinder is found to contain an explosive mixture; if they remained unmixed the hydrogen would burn quietly and the oxygen would increase the combustion of the light applied to it.

Yet another example. In a glass jar filled with air there is placed a piece of red litmus paper.

Fig. 33. Downward Diffusion of Ammonia.

Contact with ammonia would render this blue. Now let a porous vessel containing a supply of ammonia be raised upon a stand within the jar. If the gas succeeds in making its way downwards to the paper, it will do so not by virtue of its weight, for the specific gravity of ammonia is little more than half that of air. It can only effect its descent by virtue of this law of diffusion or inter-miscibility as it might be termed. Yet after a few minutes the light ammonia declares its presence at the bottom of the jar (fig. 33). The red litmus has become blue. The diffusive power of the gas has prevailed against gravity itself. Thus is shown the truth of Dalton's observations. Graham, who submitted such phenomena to a long and close investigation, discovered that the rate of diffusion of two gases is in the inverse proportion of the square roots of their relative weights or specific gravities; as, for example, in the case of oxygen and hydrogen, with specific gravities 16 and 1 respectively, as the rate of diffusion is in *inverse ratio* to the specific gravities, the gas with the higher number will have the lower rate of diffusion, but the proportion is as the *square root*; now the square root of 16 is 4, so that we arrive at the conclusion that hydrogen will diffuse four times quicker than oxygen. Or if the time of diffusion be equal, four times as much hydrogen will diffuse out of a vessel as oxygen will diffuse in.[1] To establish this fact by experiment, a tube is filled with hydrogen. In order to prevent the hydrogen escaping merely by its lightness, the upper end of the tube is closed with a plug of plaster of Paris. Now immerse the lower end

[1] Rate of diffusion of hydrogen $\left.\right\}$: $\left\{\right.$ Rate of diffusion of oxygen $\left.\right\}$ :: $\sqrt{16}$: $\sqrt{1}$.

H

in water, and leave the plaster plug exposed to
the air : the hydrogen will
presently diffuse itself outwards more rapidly than air
can enter, consequently the
liquid is drawn up into the
tube (fig. 34). In order
to be satisfied that the
lightness of the hydrogen
is not the cause of its
escape, the same experi-
ment is arranged in a different way (fig. 35). A
bottle with a plaster bottom
is filled with the gas, and a curved tube proceeding
from the neck dips downwards into water. By taking

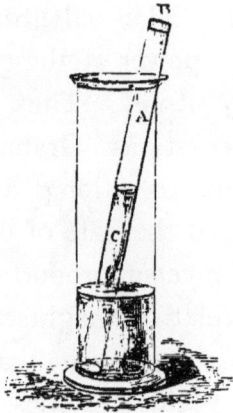

Fig. 34.   Upward Diffusion
of Hydrogen.

off the air-tight india-
rubber cap which covers
the plaster bottom you
have evidence that the
hydrogen is escaping just
as rapidly as before, by
reason of the water as-
cending the tube. Here
lightness of the gas can-
not have influence, as

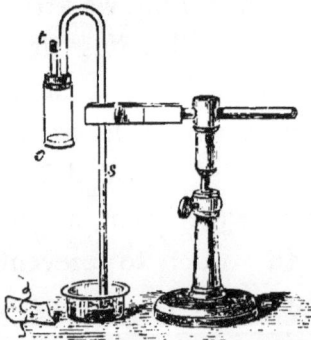

Fig. 35.   Downward Diffusion
of Hydrogen.

the escape is downwards instead of upwards. Yet
another example of the same phenomenon may

be arranged (fig. 36). On a tube 5 feet long is fastened, by means of an india-rubber ring, a porous cylinder: the lower end of the tube dips under water. By bringing a jar full of hydrogen just over the cylinder, it will be observed that air is driven out of the apparatus, and through the water; by removing the bell-jar the water rises in the tube, showing that air escapes by the porous pot. In the first case the hydrogen diffused itself into the porous vessel so much more rapidly than the air diffused itself out, that a pressure was created within which expelled some of the gas through the water; but on removing the atmosphere of hydrogen, this gas

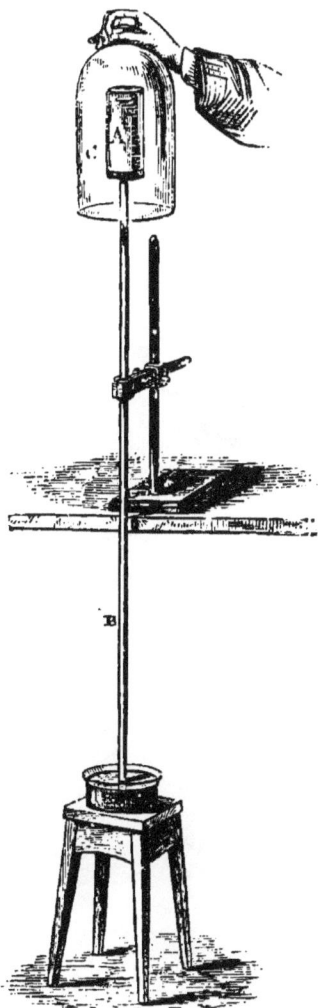

Fig. 36. Inward and Outward Diffusion of Hydrogen.

diffuses so much more quickly from the cylinder

into the surrounding air than the air can take its place, that the water is drawn up the tube. That the pressure exerted is pretty considerable can be shown, but by another form of apparatus in which the long tube passes into a two-necked bottle (fig. 37), and in the other neck of the bottle is fitted a

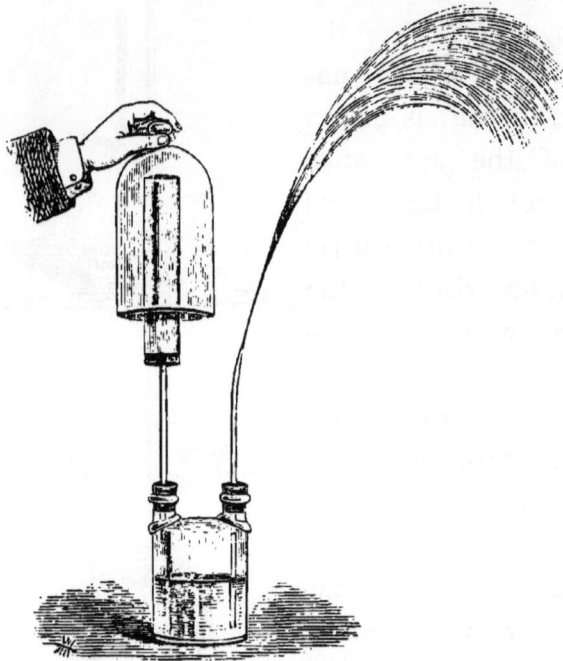

Fig. 37.   A Diffusion Fountain.

glass jet dipping below some water contained within; on holding the hydrogen jar over the cylinder a jet of water two feet in height is projected upwards, and then on letting air have access to the cylinder the hydrogen diffuses outwards, the fountain ceases to play, and air rushes violently through the liquid in the bottle to take its place in the porous cell.

The cause of these extraordinary phenomena is generally explained now by reference to molecular motion, or a constant movement or commotion amongst the particles of the gas. The kinetic or motion theory has been elaborated by Professors Clausius of Bonn and Clerk Maxwell of Cambridge. We have the best reasons for believing that equal volumes of gases at the same temperature and pressure contain the same number of molecules or particles, which molecules or particles are all of the same size. They exert their pressure in all directions, upwards as well as downwards, and thus it is that the pressure of the atmosphere is not felt by us under ordinary circumstances. If, then, a gas be under a pressure of 15lbs. on the square inch, this pressure is exerted by the gas in every direction, perpendicularly and laterally.

Two pint bottles, the one filled with hydrogen, the other with oxygen, are standing inverted, with their mouths immersed in water; the oxygen weighs 16 times as much as the hydrogen, yet the pressure in both is the same. If the particles of the gases be in a constant commotion and a hydrogen molecule strikes the vessel with only $\frac{1}{16}$ the force of the oxygen, the pressure must be equalised by many more blows being given in the same space of time, for the force with which a projectile strikes an obstacle is ascertained by multiplying its weight by the velocity with which it travels. Thus a weight of 16 lbs. travelling 1 foot per second would deliver a blow with a force of 16 lbs.; and a weight of 1 lb. at a speed of 16 feet per second would equalise this. But it must be remembered that if each hydrogen molecule hit the sides of the vessel with 16 times the velocity of the oxygen, or what is the same thing, if each hydrogen particle takes only $\frac{1}{16}$ the time of the oxygen to deliver its blow, it follows that 16 particles of hydrogen would strike in the time required by one particle of oxygen.

The pressures under such circumstances might erroneously be expressed as follows :—

	Weight		Velocity		Strokes in same time		Pressure
Oxygen . .	16	×	1	×	1	=	16
Hydrogen .	1	×	16	×	16	=	16 × 16

or the mechanical power expended by the hydrogen under these circumstances would be 16 times greater than that of the oxygen. If, however, we reduce the velocity of the hydrogen molecules to 4 times that of the oxygen, then we find—

	Weight		Velocity		Strokes in same time		Pressure
Oxygen .	16	×	1	×	1	=	16
Hydrogen .	1	×	4	×	4	=	16

This is a rational explanation of Graham's law, *'The rates of diffusion of gases are inversely proportional to the square roots of their relative weights.'*

Feddersen of Leipzig has made some interesting experiments on diffusion when influenced by temperature. His method of experimenting was as follows:—A porous stopper was inserted in a glass tube. The tube was placed horizontally, and the two ends projecting beyond the stopper were connected air-tight, by means of a caoutchouc, with other glass tubes, also placed horizontally, in each of which a drop of liquid formed a moveable stopper. Thus every propulsive movement of the column of air in the middle glass tube would cause a movement of the drops in the same direction. One end of the stopper was subjected to a constant source of heat, while the other remained cold or was artificially cooled : whereupon a slow passage of the column of air through the stopper was uniformly observed, having the direction from the cold to the warm end of the stopper. From the results obtained it appears to be a common property of porous substances, when acting as a diaphragm in gas, to occasion a passage of gas from the cold to the warm side. This phenomena of diffusion, unlike the ordinary diffusion, takes place when on both sides the porous partition there is the same gas with the same pressure ; it is proposed to give it the name of *thermo-diffusion.*

Having thus given an account of the manner in which the atmosphere is variously affected, it will now be better to proceed to the consideration of the air of close places, the composition of this under various conditions, and the possibility of keeping it similar to that of fresh air. It has already been shown that the atmosphere contains even in the open country 3 volumes of carbonic acid in 10,000 of air; that this amounts to as much as 4 or even 5 volumes under ordinary circumstances in towns. It is not advisable to breathe air containing more than 6 volumes of carbonic acid per 10,000, and, indeed, such air is pronounced by those who enter it from the outside as close, that is to say, it is disagreeable—its impurity is detected by the nose—our senses warn us not to breathe it. We cannot, therefore, conveniently increase the amount of carbonic acid in buildings by more than 2 volumes per 10,000 over that of the external air. If the amount be increased to 250 volumes the air would cease to support combustion, and 300 volumes would speedily extinguish human life, notwithstanding that 17 to 18½ per cent. of oxygen would be present, that is, 1,700 or 1,800 volumes of oxygen in 10,000. The amount of carbonic acid in the breath is about 5 per cent., so that air once breathed is useless and poisonous so far as mankind is concerned. This may be easily shown. A bell-jar (A, fig. 38), sunk an inch or two in water, but filled with air, has a cork and flexible tube B, so arranged that one may

Fig. 38. An Experiment with Respired Air.

replenish the lungs from this supply after they have been purposely emptied. After again expelling the air from the lungs into the apparatus, if a lighted taper be introduced it is instantly extinguished, and on again taking this air into the lungs a suffocating sensation will be felt. A little lighted phosphorus, however, continues to burn in the confined space, showing that all the oxygen is not used up in the act of respiration. The amount of carbonic acid given off by a man in an hour is six to seven-tenths of a cubic foot, or 14 to 16 cubic feet in the twenty-four hours. Two sperm or paraffin candles have the same effect as one man, that is to say, a good candle yields three-tenths of a cubic foot of carbonic acid in an hour. One cubic foot of gas produces two cubic feet per hour of carbonic acid, and a good gas-burner will consume three cubic feet of gas in that time. A good oil lamp gives a little more than half a cubic foot of carbonic acid gas per hour. Dr. Odling has ascertained that for equal illuminating power candles yield a larger amount of impurity to the air than gas. From these data it is readily calculated [1] that a man seated in a room with two lighted candles or a good oil lamp would require the use of 6,000 cubic feet of air in an hour in order to keep the atmosphere from becoming uncomfortably vitiated, that is, prevent the carbonic acid from rising above 6 volumes in 10,000.

Generally speaking, in dwelling-houses this is accomplished by having lofty and spacious rooms, or, in other words, having a large amount of pure air to commence with.

Another cause leading to the same end is the discomfort produced by heat, necessitating an ad-

---

[1] The carbonic acid from the lamp and the man together amounts to $1\frac{1}{5}$ cubic foot, the excess of this gas over that in atmospheric air which is allowable is 1 in 5,000 cubic feet ; if therefore we multiply $1\frac{1}{5}$ by 5,000 we arrive at the amount of air required to dilute the products of respiration and combustion.

justment of doors and windows ; this applies more
particularly to apartments of smaller dimensions.
According to Angus Smith, from experiments
made in a leaden chamber, the rise in temperature
of 170 cubic feet of air in one hour—

By the bodily heat of one man $= 5°·6$ F.

By the combustion of a candle $= 3°·8$ F.

Accordingly, in a room 8 feet high, 6 feet long,
and 4 feet broad, a man burning two candles
would in an hour raise the temperature from 60°
F. to 73° F., which alone would be sufficient to
produce uneasiness, which would be aggravated by
the physiological action of the carbonic acid present. He would in consequence of the uncom-
·fortable feeling open a door or window to get rid
of the heat and thus supply himself with fresh air.
But the heating effect of the body and of a candle
is much greater than is here indicated. Such an
experiment is imperfect because it must necessarily
last for a considerable time, and in the interval
the metallic walls are parting with the heat they
absorb from the contained air.

The amount of heat given off by the combustion
of a candle has been accurately determined by
other processes, and it has been found that a good
candle, producing $0·31$ cubic foot of carbonic acid
per hour, will raise 150 cubic feet of air $118°·4$ F.
The bodily heat of a man, calculated from the

amount of carbon consumed internally, which is indicated by the carbonic acid in the breath, will effect a rise of 139°·6 F. in the temperature of 150 cubic feet of air. This latter number is no doubt too small, because there is omitted the amount of heat resulting from the oxidation of hydrogen to water in the body, a process which must take place to some extent in the destruction of the tissues.

In default of more accurate data concerning this matter the following calculation is of interest : If the heating effect of a candle per hour on 170 cubic feet of air as found by experiment in the leaden chamber is 3°·8 F., and that of a man's body 5°·6 F., in a small quantity of air these numbers would be greater, and calculation shows that for 150 cubic feet they would be 4°·3 F. and 6°·3 F. respectively.

But the actual heating effect of a candle amounts to 118°·4 F.—hence $\frac{118°·4}{4°·3} = 27·5$, or the quantity of heat observed in the leaden chamber, must be multiplied by 27·5 to obtain the actual quantity. As the heating effect of a man's body is not widely different from this, we may apply the same calculation, so that $6°·3 \times 27·5 = 173°·2$: a number somewhat higher, and probably more correct, than that deduced from the carbonic acid in the breath, namely, 139°·6 F. These figures are instructive, showing as they do the great heating effect of a mass of human beings in a close apartment where there is an insufficient circulation of air around them to carry off the heat with sufficient rapidity.

Under ordinary circumstances the rise of temperature alone is a self-acting medium of ventilation, as already shown. It causes, firstly, a diffusion of cold air into a warmer space, and secondly, an ascension of a heavy gas like carbonic acid to the upper part of a building. A very

Fig. 39. The Ascension of Warm Carbonic Acid.

good illustration of this latter fact is afforded by the following experiment :— Two jars, holding each about half a gallon of carbonic acid, are tested with a lighted taper, which each is found to extinguish. Into each jar is lowered a little flask, one containing hot water, the other cold (fig. 39) ; after a few minutes the jar with the flask

of cold water will be found to contain sufficient carbonic acid to extinguish a taper, while
the air of the other jar will sustain combustion
brilliantly.

Furthermore, to perform the experiment on a
larger scale, take a glass vessel holding ten gallons
of the gas ; its presence within is made evident by
the floating of soap-bubbles on the surface. Now
introduce a quart flask containing warm water, and

 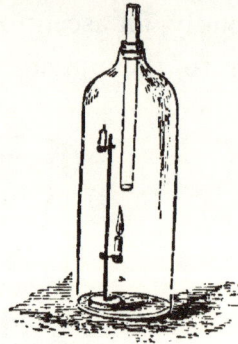

Fig. 40.   Fig. 41.

after a short interval it will be shown that the
carbonic acid has escaped by the fact of the soap-
bubbles no longer floating, but sinking to the bottom of the vessel.

All the foulest air of a room is near the ceiling ;
in fact, it is so bad there that unless an easy outlet
be provided it becomes perfectly poisonous. A
few simple experiments will illustrate the principles

of ventilation in buildings and mines in a more convincing way than mere description.

In a tall bell-glass (fig. 40) is arranged a tree of lighted candles; the carbonic acid from all of them rises, the consequence being that the top one goes out first, the second next, and so on in succession, as the gas accumulates and fills up the space from the top downwards; but if before the last and lowest is extinguished the stopper be withdrawn and the jar raised half an inch, the heated vitiated air escapes, fresh air enters, perfect ventilation results, and the light continues to burn brightly.

A similar jar has a glass chimney fitted into the neck (fig. 41); one taper is raised on the stand to the upper part of the jar, and one is placed below the chimney, and the lower rim of the jar is raised a little from the taper. The upper candle will go out, while the lower one gets a fresh supply of air from below, and discharging its products of combustion by the chimney continues to burn in the usual manner. In mines, where fresh air has to be carried downwards, the burning of fires is resorted to as a means of introducing fresh air. The principle can be well shown in the following manner:—A small box has a glass chimney fitted into each end (fig. 42). The one distinguished by the letter B has a taper suspended within it; this continues to burn brightly, the entrance of

fresh air being downwards by the shaft or chimney
A. A piece of smoking brown paper shows this:
the smoke pours down the one shaft and passing
the candle rises out of the other. Now by closing
A with a small glass plate the entrance of fresh air
downwards is prevented, and the candle is extinguished. On repeating the experiment so far, and
just before the extinction of the taper, if the glass
plate be removed the taper is restored to life again.

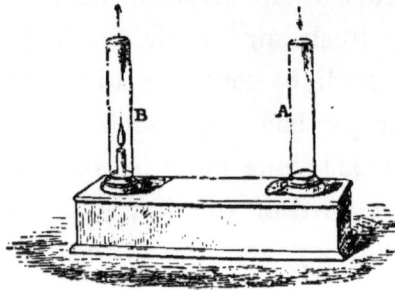

Fig. 42.

But in mines it is not often convenient to have two
shafts, and ventilation is accomplished by dividing
one by a partition. Thus in a bell-jar closed below by water, if there be fitted a glass chimney
a candle will burn until the air is so vitiated that
it is just about to die out; now on inserting a
piece of metal so as to divide the chimney into
two halves the light is revived (fig. 43), and the
smoke from brown paper will give evidence of there
being two draughts, one in an opposite direction

to the other. The smoke descends on one side of the partition, curls round in the jar and makes its exit by the other side. The causes determining which shall be the upcast and which the downcast shaft in this case are so small, the candle being as nearly as possible between the two, that we cannot detect them. In rooms the chimney is always a most important ventilating shaft, while accidental openings in doors and windows, mere chinks or cracks as they may be,

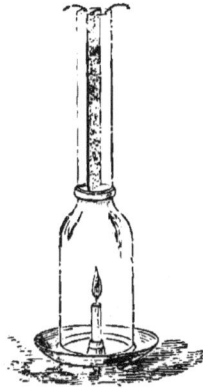

Fig. 43.

yield a large supply of fresh air. There is, however, a source of ventilation quite as efficient as the opening of doors and windows, or the accidental openings that occur in buildings ; it is a self-regulating supply of fresh air provided by diffusion through the porous walls. We are unconscious of its operation, and though the amount of air supplied by passage through bricks or stone, even when plastered and papered, is very considerable, yet one cannot detect it by any sensation of draught.

Here is the value of this means of ventilation. In winter, when it would be uncomfortable to open windows, but when, having doors and windows closed, and lamp or gas-light in our rooms, there is

I

really more urgent need of fresh air, then this unfelt and unsuspected supply is greatest. An account of an experiment of Pettenkofer's will give the best evidence of the value of penetration through the walls, and the influence of temperature on this diffusion. An ordinary room in his house with a German stove instead of an open English fire-place, the walls being of brick, and the capacity being 2,650 cubic feet, had its entire contents changed once in an hour, when the difference between the inside and outside temperature was 34° F. (66° inside, 32° outside), the doors and windows being shut. With the same difference of temperature, but with the addition of a good fire in the stove, the change of air rose to 3,320 cubic feet per hour. Now here is the most important fact : when all openings and crevices in doors and windows were pasted up and rendered thoroughly air-tight, there was still a change amounting to 1,060 cubic feet per hour, which was owing to diffusion through the walls. On lessening the difference between the internal and external temperature to only 7° F. (71° inside, 64° out), the change of air was reduced to only 780 cubic feet per hour, but on opening a window 8 feet square the change rose to 1,060 cubic feet per hour. These numbers are most interesting, showing as they do that houses may be better ventilated in winter with doors and windows shut than in summer with windows wide open, because it is

here proved that a difference of temperature of 34°
with all openings carefully closed is of greater influence, effecting a more complete change of air
than large and uninterrupted communications with
the outer air at a small difference of temperature.[1]
Hence a crowded ball-room in summer with windows wide open has a more unwholesome atmosphere than one would at first be led to suppose.
With a roaring fire the room was ventilated by the
draught of the stove to the extent of only 700
cubic feet.   The free wall of a room on being
examined for its ventilating power gave the
following numbers :—The room contained 2,650
cubic feet, and with a difference between external
and internal temperatures of 4° F., the spontaneous
ventilation or diffusion through each square yard
amounted to no less than 7 cubic feet or 43 gallons
per hour.   It has been found by Märker and
Schultze that the spontaneous passage of air
through 1 square yard of wall, the difference of
temperature being 4° F., varies with different
materials according to the following table :—

	Per hour	
Sandstone	4·7	cubic feet
Quarried limestone . . .	6·5	,,
Brick	7·9	,,
Tufaceous limestone . . .	10·1	,,
Mud	15·4	,,

[1] It is the custom in some parts of Germany for the plastered
walls to be coated thinly with paint instead of paper.   It is
doubtful which material was used in the case of this room.

The extent of free wall surface is therefore an important matter as regards the ventilation of a house ; the larger the proportion of external wall surface to the number of inhabitants the better will be the ventilation. The inhabitants of the lofty houses in the splendidly proportioned streets of Edinburgh and Paris, where families live on floors or flats, cannot have the advantage of so large a supply of fresh air through the walls in proportion to their numbers as those who dwell in the small suburban villas around London. No doubt the much lighter style of building has much to do with the matter, but setting this aside, the circulation of the wind round a house has a great effect on its ventilation no doubt, in causing a transpiration through the walls, whether through the porous materials or through chinks and crevices, and such circulation is reduced to a minimum if the houses are many-storied and constitute fine regular streets.

If dwelling-houses are not over-crowded, there is no occasion to resort to artificial ventilation. The natural ventilation caused by difference of temperature, facilitating the passage of air through the dry and porous walls, together with the assistance of windows and doors, should be sufficient, provided the greatest cleanliness throughout, and avoidance of pollution of the air is maintained. It is a curious fact, as Pettenkofer has pointed out, that the more impervious a building stone is to air the larger is the quantity of mortar generally used with it. Mortar which is made of sand and lime is, when dry, exceedingly porous, we therefore get a compensation in the larger quantity used for the lesser porosity of the stone used with it. Thus sandstones are generally so porous that water and air readily permeate them. Solid or quarried limestones are most impervious to air ; but as they are very irregular in shape, they require more mortar, and such is the reason why a wall built of such material does not differ much from a wall made of regular bricks and thin layers of mortar. Observations have shown that a wall of quarried limestone would consist of one-third mortar, tufaceous limestone one-fourth, bricks one-fifth to one-sixth, sandstone one-sixth to one-eighth.

There is a style of building now coming into vogue which consists of putting up an iron framework, which is filled with concrete ; when this material has hardened, the iron supports are taken away, and a very durable wall remains. Practical experience has yet to prove whether this material is sufficiently porous to constitute the walls of healthy houses. Pettenkofer mentions the fact that a workman's house was built near some iron-works, from large regularly shaped pieces of slag, which may best be described as an opaque

glassy substance, not in the slightest degree porous.
When completed the house dried rapidly, but when
the family had lived in it a short time the damp
became excessive and remained, thus making this
the worst house on the establishment. Moisture
condensed on the walls just as it does on the glass
of a greenhouse. To keep a house dry the walls
should be thick and porous, and the rooms well
warmed. When damp shows itself on the walls of
a house after a long spell of wet weather, it is no
indication of wet having come in from outside, but
it is due to a condition of things similar to that
causing the house built of slag to become damp
after it had been inhabited. The walls, when
thoroughly wetted, are no longer porous ; they are
also better conductors of heat, and are consequently
constantly parting with the warmth derived from
the room to the outside air, which process of
course keeps them cold. The aqueous vapour with
which the warm air of the house is saturated is
thereby condensed on the walls, and makes itself
evident to the eye.

A remarkable case in a London house has come
to my knowledge, which gives a distinct proof of
the much greater passage of gas through the walls
in winter than summer. A small room occasionally
used was noticed sometimes to have an unbearably
bad smell ; this was never noticed in summer nor

in winter unless a fire was lighted in the room : the
drainage was suspected and examined, but was
found perfect, yet here was this extraordinarily foul
air making its way into the room whenever the
interior was warm and the exterior cold.   The
cause was a dust-bin built against one of the walls,
and the filtration of the air through this and the
house wall into the room.

Having thus mentioned the causes constantly
in operation which effect a change of air within our
dwellings, it is necessary to lay emphasis on the
fact that the reckless opening of windows does not
constitute good ventilation, though it may result in
the supply of fresh air.   Those motions of the air
which are so gentle as to be unperceived by us,
causing it to stream upwards from our bodies
and through the walls of houses, must be imitated
to secure good ventilation ; that is, the supply of
fresh air *must be free from draught.*   If the rate of
an almost imperceptible breeze be measured, it is
sure to be found travelling at a higher speed than
19 inches per second.   If the hand be passed
through the air so that a space of 18 inches may be
traversed in a second, about equal to 1 mile an hour,
no draught or current of air will be felt.   The effect
is the same if the hand be still and the air moving
at this rate.   Everyone knows what a draught is
from having experienced the discomfort, if nothing

worse, of the chilling sensation caused by its continued action ; good ventilation does not admit of
draughts, generally speaking. Ventilation in tem-
perate climates is the art of supplying fresh air at
a rate not greater than 18 or 19 inches per second,
and in sufficient quantity to reduce the proportion
of carbonic acid present to 6 volumes in 10,000 of
air.   In India, where the external temperature is
so high that punkahs to create a draught are a
necessity, an essential point in ventilation is the
supply of air at a much greater speed than in
this country would be comfortable.

A person does not inhale and exhale more than
18 cubic feet of air per hour, but the supply of fresh
air necessary, according to General Morin's experiments made in Paris, is not less than 2,120 cubic
feet per hour.   I have already made a calculation
which will answer either for two men or a man and
two candles, showing that in a small apartment of
170 cubic feet capacity 6,000 cubic feet of fresh
air would be required, or 3,000 cubic feet per head.
This is the amount fixed on by Drs. Parkes and
De Chaumont.   In addition to this a very ordinary
gas-burner consuming 3 feet of gas per hour neces-
sitates a supply of 5,400 cubic feet of fresh air. Generally speaking, in this country, the change of air in
a room cannot be effected oftener than three or four
times in the course of an hour without a sensation

of draught ; hence, as 3,000 feet of air are required
for each individual, a space of 750 to 1,000 cubic
feet per head must be provided in an apartment.
The following is a practical example of the kind of
calculation necessary to form an opinion as to the
ventilation of an apartment :—A bed-room 12½
feet square and 10 feet high had an air capacity,
after deducting for the space occupied by furni-
ture, of 1,500 cubic feet ; while one wall communicated with a dressing-room of 560 cubic
feet capacity, two of the other walls were against
the open air ; there was free communication
between the room and the chimney, but both
windows and door were shut. The room had one
occupant for 9 hours, and during that time the
carbonic acid had increased only 1 volume in
10,000 of air. Taking the carbonic acid expired
during the night as 5·4 cubic feet, the air which had
passed through the room was 54,000 cubic feet ;
that is to say, the entire contents of the room had
been changed four times in every hour without the
help of open windows.

There is a ready means of ventilating a room
with an ordinarily-constructed window which is
worth describing. It consists simply in raising the
lower sash, and placing inside the window-frame a
piece of wood long enough to reach from one side
to the other, about three or four inches in height,

and an inch or so in thickness ; the window is then closed upon this so far as the breadth of the wood will allow. The inside sash frame is thus raised three or four inches above the place where it should meet the outer one, and as there is an open space between the two sets of window-panes, a current of air moves inwards and upwards towards the ceiling of the room, and thus no draught, will be felt. This admirable arrangement, which is at least twenty years old, answers well in small bed-rooms where the window cannot be opened without causing a cold draught, and it is so simple and costless that those who may be doubtful of the comfort it will afford, can have no excuse for not giving it a trial. There is yet another method, which consists in perforating with perpendicular holes the thick wood of the frame where the two sashes meet. The excellent plan which has been advocated by Mr. Tobin for use in private houses, differs from this only by cutting the wood away on each side, right and left of the spring fastener. In case at any time the supply of air should be too great through such a wide opening, a flap or lid of sheet zinc is hinged on, so as to close it when desirable.

The ventilation of ordinary dwellings is a simple matter ; but in considering the case of public buildings and workshops, where the number of people crowded together is very great, and the amount of gas consumed extremely large, there is under such circumstances a state of things almost beyond the control of natural ventilation. Fresh air should be provided by well devised means, and the greatest possible care should be taken that natural ventilation should have full play ; but how seldom this is the case the following tables will show :—

Carbonic Acid in close places in London, according to Dr. Angus Smith :—

1864	Parts in 10,000 of Air
Chancery Court, 7 ft. from ground, closed doors, March 3	19·3
Same, 3 ft. from ground	20·3
,, door wide open	5·
Strand Theatre, gallery, 10 P.M.	10·1
Surrey Theatre, boxes, ,, ,,	11·1
,, ,, 12 ,,	21·8
Olympic, 11.30 P.M.	8·17
,, 11·55 ,,	10·14
Victoria Theatre, boxes, 10 P.M.	12·6
Haymarket, dress circle, 11.30 P.M.	7·57
Victoria Theatre, boxes, April 4	7·6
City of London Theatre, pit, 11.15 P.M.	25·2
Standard Theatre, pit, 11 P.M.	32·0
St. Thomas's Hospital, Queen's ward, 3.25 P.M.	4·0
,, ,, Edward's ward, 3.30 P.M.	5·2
Lambeth Workhouse wards	·1
St. Luke's (Chelsea)	7·6
East London (Homerton)	7·6

The enormous volume of air necessary to preserve a wholesome atmosphere in a theatre may be readily understood when it is known that Drury Lane will hold an audience of 4,500 persons.

Taking General Morin's minimum amount of 2,120 cubic feet of air per hour for each person, we find that more than nine and a half millions of cubic feet of air per hour must be provided for a crowded audience.   In this calculation no account is taken of the immense number of gas-lights using up oxygen and pouring out impurity ; it does not therefore exaggerate the matter.

In April 1874 I made experiments to ascertain the amount of carbonic acid in the air in Covent Garden and Drury Lane theatres during the performances of Italian opera, and also in the Money Order Office in Aldersgate Street.   In the theatres it was considered, from the amount of gas burnt and the small difference between the outside and inside temperature, besides the fact that no extensive provision for ventilation was in use, that the air would be very bad, much worse than in winter, when the outside temperature is low.   In the theatres the escape of burnt gas is not sufficiently provided for, and the products of combustion strike the roof and roll over into the galleries.   This arises not so much from want of outlet as for lack of inlet of fresh air to displace that rendered hurtful by heat and impurities.   Public buildings too often resemble the bell-jar on page 113, fig. 43, in which, though a candle is burning beneath an ample opening, the vitiated air does not escape for want of sufficient inward draught. But as soon as the chimney is divided, upward and downward currents are established and ventilation ensues. If shafts were carried downwards from the roof to the basement of a theatre there can be no doubt similar effects would follow. The gallery and boxes, too, of a theatre may be likened to a number of small and over-crowded apartments, being as they are so shut in overhead and at the back. That this is the case anyone may prove for himself by passing the hand up from floor to ceiling when, between the acts, the door is open on a warm night. The cold air rushes in at the bottom, and at the top the hot air escapes : two currents are playing in opposite directions, one over the other.

Analyses made in April 1874.

	Volumes of Carbonic Acid in 10,000 of Air
Covent Garden Theatre, amphitheatre . . .	22·0
,, ,, near an open door, time 10.10	17·6
,, ,, near a ventilator, time 10.20 .	14·8
Drury Lane—in the lower gallery	27·0
Money Order Office	17·4

Here it is easily seen that the air in the theatres is very bad ; but after the doors had been open for a short time between the acts it rapidly improved ; indeed, in Covent Garden in the second case, near an open door, the people exclaimed, by force of

contrast, how delightful the fresh air was; nevertheless, this *fresh air* contained 14·8 parts of carbonic
acid in 10,000, or from $2\frac{1}{2}$ to 3 times as much as
it should have had.   Drury Lane was the first
place experimented on, and as I had entered at the
commencement of the performance, the bad effect
of the air as it became vitiated was only gradually
experienced, but it produced a listlessness and
headache.   All the audience around were evidently
affected in the same manner, and appeared to be
constantly sighing and gaping, or in other words
gasping for breath.   In the upper boxes of the
same house, a rather more confined space, it seemed
as if the breath of those in front was passed on
from mouth to mouth, the last recipients being
those seated farthest behind.   On entering Covent
Garden from the street in the middle of the performance it was scarcely possible to remain in
the building for more than twenty minutes, and the
feeling of relief on coming out again was immense.
The effect experienced by many persons after
breathing such foul air for three or four hours results in a feverish state and headache, even after
so long an interval as twelve or twenty-four hours.
The immediate effect is to weaken and decrease the
rate of the pulse, and at the same time quicken the
respiration.   There can be no question that the
effect of the performance and the enjoyment of the

audience are very seriously damaged by the listlessness and dulness of comprehension and feeling
induced—a condition which is only occasionally
dispelled for a time by some increased exertion on
the part of the performers.

The bad effect of carbonic acid in the air alone,
without taking into account organic matter, has
been shown by Dr. Angus Smith, who ascertained
that 1 part of the gas in 1,000 of air produced in
15 minutes an increase in the number of respirations
from 18 to 19 per minute, which increase remained
the same up to 45 minutes ; the pulse was lowered
in 25 minutes from 73 to 71 beats, while at 45
minutes it was 72 per minute. With $2\frac{1}{2}$ volumes of
carbonic acid in 1,000 of air the pulse, at first 70
beats per minute, increased to 73 at the end of 10
minutes, and at the end of half an hour was lowered
to 69, while the inspiration increased from 17 to 21
per minute. With 5 volumes of carbonic acid per
1,000, the pulse at first 76, and the inspiration at
17, at the end of 40 minutes were represented by
the numbers 71 and 24.

Passing over extreme cases of poisoning by
foul air, such as the well-known one of the Black
Hole of Calcutta, and of the prison in which 260
out of 300 Austrian prisoners died after the battle
of Austerlitz, it is necessary to lay great stress
upon the now well-ascertained fact that air only

moderately vitiated, if breathed for a long time day after day, produces most serious results. The effects are seen in pale faces, partial loss of appetite, a lowering of the spirits, and a decrease of muscular strength. Very frequently such affections of the respiratory organs as bronchitis and phthisis result. Dr. Parkes, in his work on 'Practical Hygiene,' mentions the fact that in the badly ventilated prison of Leopoldstadt in Vienna, in the years 1834 to 1847, the proportion of deaths was 86 per 1,000, out of which number 51·4 per 1,000 were due to phthisis, while in the well-ventilated House of Correction of the same city the deaths were 14 per 1,000, of which 7·9 were occasioned by phthisis ; 43·5 cases of phthisis may be therefore attributed to foul air. The origin of phthisis and other lung diseases may reasonably be considered due to the inhalation of organic particles thrown off from the lungs of unhealthy persons, and it is now a well-ascertained fact that a bad atmosphere assists the rapid spread of several specific diseases, such as typhus, small-pox, and scarlet fever. From personal observations convincing evidence was obtained that such was the case with the cattle plague in 1866. In cattle sheds containing twenty or thirty cows—which the owners kept closed to such an extent that all chinks in the doors and windows were stuffed with straw and matting, under an ignorant belief that thus the plague could be kept out—very frequently the entire stock died in two or three days after the first appearance of disease ; while in other cases, where the animals were housed in a well-cleaned and tidily-kept shed, with a plentiful supply of fresh air, not only did some of them escape the disease altogether, but the deaths were reduced to one-third of the number of beasts attacked.

The large supply of air necessary in hospitals for contagious diseases is fully recognised by medical men, and more especially so in America.

The next matter—of greater importance than it may at first sight appear—is the composition of the air permeating the soil. Boussingault and Lewy, in 1853, first made researches in this direction, and found that this air is very rich in carbonic acid, no matter whether the soil be sandy, loamy, or rich in organic matter. The amount of carbonic acid is much greater than that of the atmosphere, though of course earth, with a large proportion of decomposing organic matter in its composition, has a still greater proportion of this gas. The following table illustrates this :—

Carbonic Acid in the Air of the Soil.

BOUSSINGAULT AND LEWY.

Description of land					Parts in 10,000
Field recently manured .	.	.	.	.	221
,, ,,	.	.	.	.	974

K

Description of land		Parts in 10,000
Field of carrots		98
A vineyard		96
Forest land		86
,, loamy subsoil . .		82
,, sandy subsoil . .		24
Garden soil		364
Prairie soil		179

The fact that a so much greater amount of carbonic acid is present in what may be called the ground air, than occurs in the respirable atmosphere, is readily shown by the following arrangement (see fig. 44). A flask, full of clear baryta

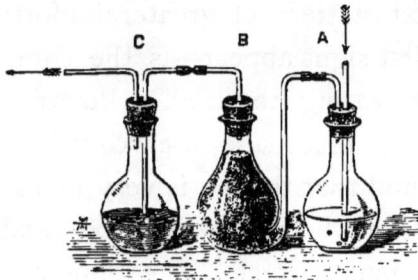

Fig. 44.   The Presence of Carbonic Acid in the Air of the
Soil shown.

water, is attached by tubes to a vessel filled with earth, and again attached to this is another flask of baryta solution ; by drawing air through the whole system of bottles the amount of insoluble carbonate of baryta produced in the first flask by the carbonic acid in the air, may be compared with that in the second flask produced by the car-

bonic acid in the soil. In fact, after a very short
time, it is made perfectly evident that earth dug
out of a meadow renders to the air passing through
it at least ten times as much carbonic acid as the
air originally contained. Pettenkofer has thought
it advisable to investigate the nature of the ground
air twice a week all the year round in Munich,
and he considers it a matter of much importance
and worthy of great attention. His ascertained
facts are surprising, and a much larger number of
experiments is necessary before we can account
for them. In sterile land, consisting of chalky
rubble, he finds the quantity of carbonic acid is
smaller at 5 feet than 13 feet throughout the
year, excepting in the months of June and
July, when an inverse proportion arises. In the
lower stratum, however, a considerable increase
soon begins again, which continues until the maxi-
mum is attained. The carbonic acid in the soil of
Dresden is far greater in amount—nearly twice as
much as in that of Munich. We cannot connect
these facts with any other knowledge which would
lead to their explanation, but they teach us an
important lesson concerning the surroundings of
our houses. The soil is very easily permeated by
the outer atmosphere—more especially if some
pressure, such as that of the wind, is brought to
bear upon it. That this is the case may be shown

thus :—A long cylinder is filled with earth, to the bottom of which goes a tube. At the upper end the tube is bent like a **U**, and a little water is placed in the bend ; now, by blowing with the mouth through a tube on to the surface of the soil, the liquid is seen to be raised, which indicates that a measureable amount of pressure was exerted. It has been shown that the bulk of a gravelly soil consists of one-third air ; that is, the space between the stones and particles of sand which is filled with air amounts to one-half the space filled by the gravel. Let us now consider this soil to be the surroundings of a house, and imagine a drain leaking into such soil ; the ground air would be charged with the products of decomposing offensive matter, and those abominations which are to be so much dreaded as the cause of zymotic diseases. The house built upon this soil is full of warm air, and up the chimneys a draught rushes, causing, when the doors and windows are closed in the lowest rooms, a certain amount of the ground air to enter by way of the floor, carrying with it the germs of disease and perhaps death. Pettenkofer mentions the fact that persons were poisoned and killed by coal gas in a house where no gas was laid on. It had travelled twenty feet under the street, and then through the foundations, cellars, and flooring of the lower rooms. A most serious explosion, which demolished at least one house, occurred in the latter part of the year 1874 at Southgate, near London. The houses had no gas laid on, but the main pipe which ran along the road was damaged, the coal gas permeated the soil, and escaped into the dwellings, where it was smelt for some days before the final disaster happened. These cases both occurred in winter, when the house being warmed acted as a chimney shaft. It shows that a badly made drain and ill-fitted gas-pipes may be the death not only of the inmates of the house to which they belong, but also of the neighbours who have no control over them. These remarks do not apply to buildings with properly ventilated cellarage, but they show the importance of placing ventilating gratings in the wall between the soil and the flooring of basement rooms. In the suburbs of London it is not at all uncommon for speculative builders to erect houses the drain-pipes of which have no proper outlet, so that until they become filled with sewage, and the surrounding soil saturated—which may take about twelve months—the unfortunate inhabitants are unaware of the fact. In the meantime the mischief so much to be dreaded may have occurred. Against such practices there is no protection at present, or, what is the same thing, there is nothing to prevent such reckless dishonesty. It is a question how far a clay soil is or is not impervious to sewer gases, and therefore to what extent the air of houses built on clay may be free from contamination with impure ground air ; but enough has been said to show how dangerous is defective drainage, and how easy is the escape of sewer gases through the soil into dwelling-houses, the earth actually forming a reservoir for them. It was pointed out some years ago that evil effects arose from living in houses built on sites originally gravelly, but from which the gravel had been removed and replaced by the refuse from dust-bins. From complaints recently made this practice seems to be still carried out. It is unpleasant to think of living on a dust-heap, and the preceding remarks have made the danger of it obvious.

CHAPTER IV.

THE facts connected with putrefaction and decay have from superficial observation led to the erroneous belief which, having originated in the earliest times, has survived to some extent until the present day—that minute organisms, the lower order of living things, spring from dead matter ; that worms, for example, are derived from putrefying flesh, that they are not reproduced from parent organisms, but are generated from decaying substances   A few quotations will illustrate this.

Lord Bacon, in the ' Novum Organum ' (Book II., Aphorism 13), says : ' All putrefaction exhibits some slight degree of heat, though not enough to

be perceptible to the touch : for neither the substances which by putrefaction are converted into
animalculæ, as flesh and cheese, nor rotten wood
which shines in the dark, are warm to the touch.'
The joke of Dr. Johnson on Tom Davies, a bank-
rupt bookseller who took to authorship, that he
was 'an author generated by the corruption of a
bookseller,' is evidently a hint as to his connection
with 'Grub Street' through an allusion to the
popular belief.

In Shakspeare's 'Antony and Cleopatra' (Act
II. Scene VIII.), the mud of the Nile is said to give
birth to serpents and crocodiles by the operation
of the sun.   In 1662 Van Helmont relates how a
green leaf placed between two bricks was transformed into a scorpion, and states that living things
—fish, newts, and frogs—in ponds that have been
dried up, spring from the mud ; he further tells
how to generate mice out of some sawdust and an
old shirt ; and in 1870 Dr. Bastian (a Fellow of
the Royal Society) relates how various microscopic
fungi may be developed from solutions containing
only definite chemical compounds.   Van Helmont
enters into an explanation of the phenomena he
describes, and attributes the cause to the presence
of an essence which impregnates the mass, fer-
tilises it, and makes it bring forth life.   This is
really a very important theoretical assumption of

Van Helmont, which has much more apparent truth in it than the facts offered in its support would lead one to believe. Most careful experiments made within the last fifteen years did not lead to any more definite result than a confirmation to some extent of this theory.

Redi in 1638 made the first discovery which undermined the belief in spontaneous generation. He found that when meat putrefied, the maggots were not generated from the flesh, but were the larvæ developed from the eggs of the common blow-fly ; that when the meat was covered with gauze the maggots did not appear, but the flies settled on the gauze nearest to the meat and there deposited their eggs. The common wire-gauze dish-cover is the practical application of Redi's discovery to household purposes. At the beginning of the eighteenth century microscopical research revealed in infusions of vegetable and animal matter a great variety of minute living things, the existence of which not being satisfactorily accounted for led the discoverers to believe that, under certain favourable conditions, matter which had once lived preserved after death a sort of vitality, under the influence of which it underwent a re-arrangement and re-construction, producing a different class of organisms. There appeared in London, in 1745, a work by Needham supporting

the doctrine of spontaneous generation by arguments derived for the first time from direct experiments.   Needham's views were adopted by Buffon,
but they had both partisans and opponents.   The
latter advocated Bonnet's theory of the pre-existence
of living germs in the atmosphere, and amongst
them the most distinguished advocate was the
Abbé Spallanzani, who, repeating Needham's
experiments, obtained opposite results.   It suf-
fices to say regarding these experiments that
vessels containing putrescible substances were hermetically sealed and heated to the temperature of
boiling water.   Needham heated his vessels for a
space of five minutes only, and they contained
organisms, while in Spallanzani's experiments they
were heated not less than forty-five minutes, and
were found barren.   Needham then complained
that Spallanzani's 'had put to the torture' and
'not only much enfeebled, if not totally annihilated
the vegetative force of the infused substances, but
also that he has totally destroyed by the exhalations and heat of the fire the small portion of air
remaining in the empty part of the flask.'   It is to
be observed that Needham has no evidence whatever to support such statements.   Unfortunately
the discovery of the composition of air by Lavoisier
had not yet been made, consequently eudiometry
had not yet been invented.   No information, there-

fore, concerning the validity of this objection was at this time obtainable.

The real value of any laboratory experiment is put to a very severe test when it is carried out on a large scale for commercial purposes. Appert applied to domestic purposes Spallanzani's results carried out according to Needham's plan, and the process has now developed into the preservation and importation of immense quantities of Australian meat, and the peas and other vegetables procurable in winter in a Parisian restaurant. Gay-Lussac reported on Appert's process of preserving food, and having found the air contained within the vessel to be destitute of oxygen, came to the conclusion that the absence of oxygen was a necessary condition of the preservation of animal and vegetable substances, a doctrine which has been very generally received, but which is in reality most erroneous. Schwann, in 1837, first showed this, and refuted the objection of Needham that Spallanzani had destroyed the air in his flasks by prolonged heating. His experiments show that fresh, cool air, if previously heated, has no effect on the juice of meat which has been boiled, and his conclusion is, that fermentation and putrefaction are not caused by oxygen, or by means of oxygen in the air, but by a principle included in ordinary air which heat can destroy. Schultze modified

Schwann's experiments by purifying the air admitted to his infusions by passing it through energetic chemical reagents, such as oil of vitriol and caustic potash. Schrœder and Dusch, guided by the experiments of Lœwel, which showed that ordinary air was not able to provoke the crystallisation of sulphate of soda when filtered through cotton wool, applied the same principle to the preservation of infusions, and by means of flasks, fitted up in the same manner as those used in Lœwel's experiments, they operated on the following materials :—

1st. Meat with the addition of water.

2nd. Beer wort.

3rd. Milk.

4th. Meat without water.

In the first two cases the air filtered through cotton wool left the liquids unchanged even after many weeks ; but the milk soon curdled and the meat putrefied. Five years afterwards, in 1859, Schrœder alone returns to this subject, and after recounting numerous and various very interesting experiments, concludes thus:—'It must be admitted that fresh air contains an active substance which provokes the phenomena of alcoholic fermentation and putrefaction, a substance which heat destroys and cotton wool arrests. Must this active substance be regarded as formed of organised microscopic germs disseminated in the air? Or is it a chemical substance still unknown? I cannot tell.'

In 1859 the many difficulties besetting this subject were attacked by M. Pasteur in a most impartial spirit, free from preconceived ideas, and with the object in view of founding a well-grounded opinion upon this important question of hetero-genesis. The conclusions of Schwann and Schrœder could leave no doubt in the mind of the reader of the existence in the air of a principle which is the necessary condition of life appearing in infusions. Those who stated that this principle was nothing but germs had no further proofs in support of such an opinion than those who thought it might be a gas or a fluid, and who consequently inclined to the belief in heterogenesis. Schwann, Schrœder, and Schultze failed when experimenting in certain liquids, and Pasteur failed constantly and with all liquids when, as will be presently noticed, his liquids came in contact with mercury, the cause of these failures being at the time quite unknown, no cause of error being even suspected.

Thus we see that in times as recent as 1859— the date of Schrœder's memoir—we had really no more definite information than that given in Van Helmont's statement in 1662, *but then it had been established by careful and exact experiments.*

About this time M. Pouchet, a renowned naturalist of Rouen, published a work advocating the theory of ' Heterogenesis.' He supposed that when water, a putrescible substance, and air are brought together, certain so-called plastic forces of matter beget germs, which germs undergo the same process of development as those resulting from the normal process. Animalculæ, he said, were formed from the débris of others which preceded them. He repeated Schwann and Schultze's experiments, and failed to get the same results. The liquid mostly used by Pouchet was an infusion of hay, and his arrangements for excluding germs from the air in his experiments were very insufficient. He objected that if the air were a reservoir for germs it would be so overloaded with organic matter as to be quite foggy, and that these germs would not escape microscopic observation. He examined dust which had settled in different places, and found it contained grains of starch, silica, soot, and spores of fungi, but the latter, he said, were very few in number.

In 1862, two hundred years after Van Helmont, Pasteur's investigation of all the principal foregoing facts was published. The ingenuity displayed and the remarkable chain of evidence afforded by these most carefully executed, most complete, and therefore, most trustworthy experiments, give the whole paper, when read in detail, more the air of a romance

than the record of scientific laboratory work. At
the same time it completely sets at rest the doubts
concerning the existence of countless numbers of
living organisms in the air, and furnishes conclusive evidence against the spontaneous generation
of living things.

*Pasteur's Microscopic Examination of the Solid
Particles diffused in the Atmosphere.*

The question which Pasteur first set himself to
answer was—Is it possible to gain an approximate
idea of the relation a volume of ordinary air bears
to the number of germs that this air may contain ?
That there are germs in the air is admitted by the
staunchest partisan of the doctrine of spontaneous
generation—M. Pouchet. He has recognised the
spores of cryptograms, and particles of starch in dust,
but he adds that their number is very small. It is,
however, not the dust deposited on surfaces such
as Pouchet took which should be examined, but
that floating in the air, for, as Pasteur remarks, the
organised particles may be considered the lightest
and would be carried about by draughts, whilst
the heavier particles would settle. This fact has
been overlooked by some who, in criticising the remark of Professor Tyndall, that floating dust consisted chiefly of organic matter, which conclusion
he came to from the fact that it was combus-

tible, said that Dr. Percy had found the dust settled on the walls of the British Museum consisted of 50 per cent. of incombustible matter. Now it is perfectly evident that, as a whole, organic matter has a lower specific gravity than inorganic, and organised organic matter may be expected to be still lighter : on this account if 50 per cent. of the dust deposited on surfaces were organic, a still higher proportion of that which floats in the air would be of the same nature, while organisms would for the most part float in ordinary air which is subject to disturbance.

Water,[1] which is more than 800 times heavier than air, remains suspended in the form of very fine spherules, causing a mist or fog when near the earth's surface, and constituting the clouds which float high above. Solid particles of carbon are discharged into the air by chimneys in the form of smoke. Taking these facts into account, we cannot be surprised at finding the dust (which is generally speaking so fine as to be unseen) to consist of such very finely divided solid matter as is most abundant around us.

But let us see by what means Pasteur sought

---

[1] One cubic foot of water weighs 1,000 oz., 13 cubic feet of air weigh 16 oz. $\frac{16}{13} = 1\cdot23$ and $\frac{1000}{1\cdot23} = 813$, the proportion by which water is heavier than air.

to determine the nature of floating particles in the air, and what conclusions he arrived at.

By means of a water aspirator (see fig. 45), he drew air from a quiet street and from the garden of the École Normale in Paris at some distance from

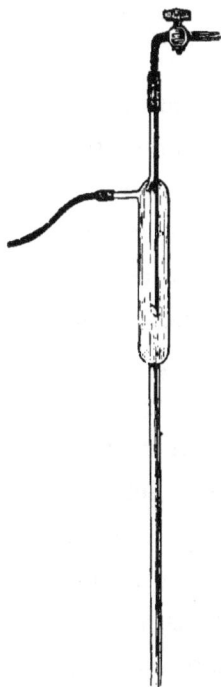

The fall of water through the wide vertical tube carries bubbles of air before it, the air being drawn through the narrow horizontal tube attached to the wide one. At the point where the water is discharged the air also escapes : this can be collected in a flask full of water inverted in another vessel. It is easy thus to measure the quantity discharged in a minute.

Fig. 45.   Water Aspirator.

the ground through a tube containing a plug not of cotton wool as in the experiments of Schrœder but of soluble pyroxyline, such as is used for making collodion. Pyroxyline is cotton wool which has been steeped in a mixture of strong nitric and

L

sulphuric acids, and afterwards carefully washed with water and dried. It is so altered in chemical properties as to be easily soluble in a mixture of strong spirits of wine and ether. It burns rapidly, and if prepared in a particular manner is converted into the highly explosive substance—gun-cotton.

By drawing air through a plug of this soluble cotton, inserted in a small glass tube, all the solid particles were intercepted or filtered off. The amount of air aspirated in a given time was accurately measured, and after a sufficient interval the soluble cotton plug was removed and treated with its solvents, alcohol and ether. After allowing the dust to subside in a tube, the collodion was syphoned off and more alcohol and ether was repeatedly added, to effect the perfect removal of the collodion. The completely washed dust was placed on a microscope slip, and examined in a drop of water. By ordinary methods the action of different reagents, such as iodine water, potash, sulphuric acid, and colouring matters, on the particles was tried.

The accompanying drawings represent some of the organised corpuscles collected during twenty-four hours on November 16 and 17, 1859. Immediately after washing the dust a small portion was placed on a slip of glass and tested with a drop of potash-solution containing 5 parts of the sub-

stance in 1.00 of water (see fig. 46). It was then
covered with a microscopic thin glass cover and

 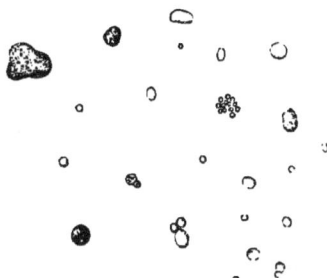

Fig. 46.   Fig. 47.

examined, and when an evidently organised particle was noticed this was drawn. The potash-
solution was then withdrawn from under the glass
by means of absorption with filter paper, and
iodine water was substituted. Fig. 47 shows the
action of iodine water, and fig. 48 the subsequent
effect produced by ordinary
concentrated sulphuric acid.

This process disclosed the
fact that there is in ordinary
air a variable number of corpuscles of such forms and
structures as show them to be
organised.

Fig. 48.

Their dimensions range from extreme
smallness up to the diameters of 0·01 millimètre
to 0·015 millimètre or more. Some are perfectly
spherical, others ovoid. Many are translucent,

L 2

others opaque, with internal granulations.   Those
translucent particles of a regular shape so closely
resembled the spores of the most common fungi
that the most able microscopist could not see any
difference.   These corpuscles were evidently organised, resembling completely the germs of the
lowest organisms, and so diverse in size and structure as to belong without doubt to very various
species.   The employment of iodine [1] shows that
amongst those particles there are always some
starch granules, but it is easy to separate all such
granules by treatment with
sulphuric acid, which dissolves
them.   Doubtless other particles also are dissolved, but
still there remain a great
number, which are often seen
more clearly by reason that
they are freed from the carbonate of lime and other particles of amorphous dust. But
by direct experiment it was
found that ordinary oil of
vitriol did not dissolve the
spores of common fungi, even after prolonged

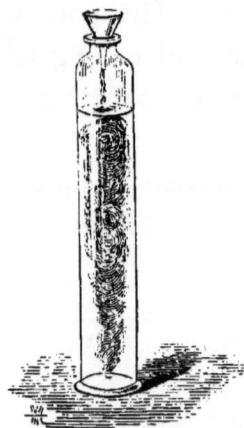

Fig. 49.   Starch
coloured with Iodine.

[1] The iodine, as before remarked (see p. 60), strikes a deep blue,
indeed inky black colour, with minute portions of starch, so that a
test-tube full of starch solution in half a gallon of water with a little
iodine, gives this intense colouration.   See fig. 49.

contact. It is needless to say that the soluble cotton used was previously tested and found to contain no residue insoluble in alcohol and ether beyond a fibre or two.

By exposing for 24 hours a plug of pyroxyline to a current of air passing at the rate of a litre the minute after a succession of fine spring days, it was found that many myriads of organised corpuscles were collected. It may be easily understood that the time of the year, and whether before or after rain, and a thousand and one causes, may increase or lessen the number of solid particles which everyone has seen dancing in a ray of sunlight entering a darkened room. Pasteur found that in the winter months, particularly once on the

Fig. 50. Dust collected on
June 25–26, 1860.

Fig. 51. Dust collected during
an intense Fog in February 1861.

occasion of so low a temperature as $-9°$ C., to $-14°$ C. (or $6°$ F., to $7°$ F.), a very small number of germs were collected.

The drawings, figs. 50 and 51, show both organised and amorphous particles, such as are seen in the field of a microscope giving an enlargement of 350 diameters, the moistening liquid being ordinary sulphuric acid.

Pasteur remarks near the end of his second chapter, ' I think it would be of great interest to multiply researches on this subject and to compare the organised corpuscles disseminated in the air at one place during different seasons and at different places at the same time. It appears that our knowledge of contagious diseases, especially at periods when epidemics rage, would be increased by work carried out in this direction.' The influence of this suggestion may be traced in recent advances in surgical and medical science. By following up his own ideas Pasteur was enabled to prescribe a means of preventing the disease known as ' pébrine,' which made such havoc amongst the silkworms in France.

As mention will have to be made of those organisms which develop in decomposable organic liquids, it is necessary to give a slight description of them. The commonest and most easily observed are the minute fungi commonly known as mould or mildew. When a previously boiled putrescible liquid is left open to the air, exposed to a warmth of about 30° C. (95° F.), there generally appears a slight speck of down upon the surface : this increases to a tuft, which on examination with a

low microscopic power not unfrequently is resolved
into an appearance like that seen in fig. 52. The
growth consists really of two differently disposed

Fig. 52. A common form of Mould.

parts, the mycelium and the fructification. The
former (fig. 53), which performs the function of

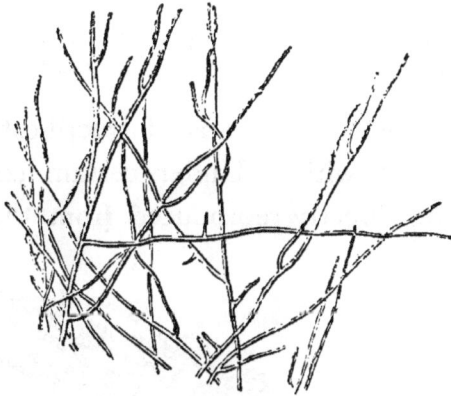

Fig. 53. Mycelium of Fungus. After Pasteur.

nutrition, is a network of tubes immersed in the
liquid. The latter, raised on fine threads, consists of
spores serving for the propagation of the species.
In the *mucorini* the threads are terminated by
sporangia (fig. 54) containing sporidia, which are set

free when mature by the bursting of the sporan-
gium (see fig. 55).

Fig. 54. Sporangia of   Fig. 55. Sporidia set free.
Mucor.   After Pasteur.   After Pasteur.

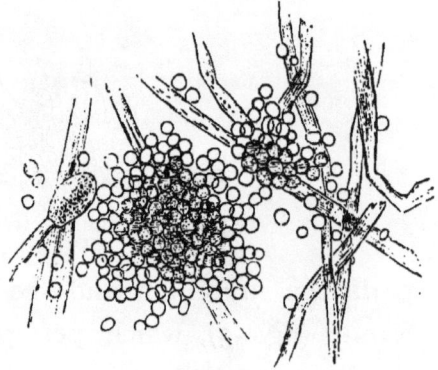

Figs. 56 and 57 are forms of fructification very
commonly met with.   Under certain conditions
fungi can be directly reproduced from the myce-

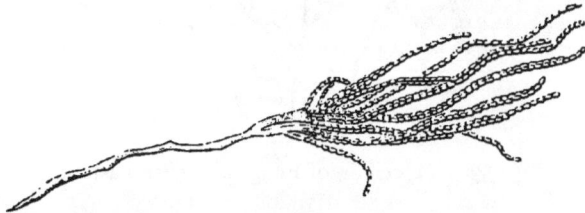

Fig. 56.   Fructification of Penicillium.   After Pasteur.

lium without spores.   There are two great divisions
of fungi, and the commonest forms of mould, referred to their family, order, and genus, are given
here—

I. *Sporifera*, spores naked. Family, *Hypho-
mycetes*. Order, *Mucedines*. Genera, *Penicillium*
(fig. 56) and *Aspergillus*.

II. *Sporidiifera*; Sporidia in vesicles. Family,
Physomycetes. Order, *Mucorini*. Genus, *Mucor*.

A very frequently occurring growth, especially
in saccharine liquids, is the torula [1] (fig. 57). This has

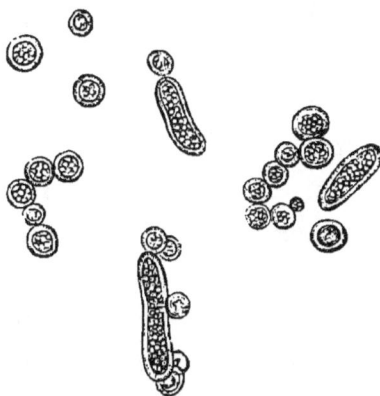

Fig. 57. Torula.

no mycelium, but consists simply of cellular organisms, which frequently elongate and branch : it
is essentially the cause of alcoholic fermentation, the conversion of sugar into alcohol. Besides
these there are certain organisms of much smaller
dimensions seemingly allied to the Algæ, which are
the constant attendants on putrefaction.

These are bacteria, and according to Cohn's

[1] According to the Rev. M. J. Berkeley, torula is but an altered
form of penicillium, but De Bary does not confirm this.

researches and his definition they are 'cells free from chlorophyll, of spherical, oblong, or cylindrical form, sometimes twisted or bent ; they multiply exclusively by transverse division and occur either isolated or in cell families.' They differ from fungi in this that such a thing as branching is never seen. They frequently form a jelly-like mass, which is called the *Zooglæa* form. Attracted by oxygen, bacteria also collect like a layer of oil upon a liquid, or as a tough pellicle, a modified *Zooglæa*, the little rod-like organisms being packed on end in rows. After a certain time they are generally to be found at the bottom of a vessel as a pulverulent precipitate ; this occurs when the liquid no longer yields them nutriment. In a perfectly clear liquid, such as a carefully prepared turnip infusion, they make their first appearance as a slight turbidity. This is really due to their incessant and rapid motion, for after they have subsided the liquid again becomes clear. Most bacteria present a motile and a motionless condition.

The following illustration gives an idea of the appearance of different organisms classed as bacteria by Cohn, though only A and D are what may be called bacteria proper. These have all been observed with one of Messrs. Beck's best one-eighth inch objectives and the second eyepiece of one

of their microscopes.   Magnifying power 750 diameters.

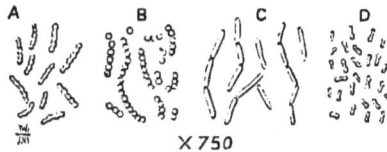

Fig. 58.   Some of the most minute Organisms found in organic
infusions, &c.

A. *Bacterium Lineola*—the *Vibrio lineola* of
Ehrenberg—commonly found in stale milk, ' Pas-
teur's solution,' and other putrefying liquids.
Motion very rapid.

B. *Micrococcus ureæ*.   The necklace fungus
supposed by Pasteur to convert urea into carbonate of ammonia.

C. *Bacillus ulna*.   Seen abundantly in perfectly
fresh hay infusion.   Jointed rods.   Motionless.

D. *Bacterium termo*.   The motile form constantly seen in early stages
of putrefaction.

Fig. 59.   Bacillus subtilis.
Vibrio lineola.   Pasteur.

The following (fig. 59),
taken from Pasteur's drawing, is what he describes
as *Vibrio lineola* ; it is the
ferment of sour milk, and
the cause of that action
which converts lactic into
butyric acid.   Cohn calls it *Bacillus subtilis*.

Pasteur's Experiments with Heated Air.

Although it appears there are in air organised corpuscles in great numbers which are indistinguishable from the germs of the lowest organisms, is it really a fact that amongst these there are particles capable of germination?   This interesting question was answered in a conclusive manner. Firstly, the facts announced by Schwann were firmly established, although they had previously been attacked by Mantegazza, Joly and Musset, and Pouchet.   The solution, sealed up in flasks, was one extremely liable to change ; its composition was —

Water	100 parts
Sugar	10 ,,
Albuminoid and mineral matters from yeast	0·2 to 0·7 part

This was boiled for two or three minutes, and then placed in contact with air previously heated to redness, and not a single doubtful result was obtained, although the experiment was repeated at least fifty times ; no trace of any organised production was seen even after the liquid had been kept eighteen months at a temperature of 77° to 86° F. ; whereas, if the liquid be exposed to ordinary air for a day or two, it never fails to become filled with bacteria or vibriones, or covered with mould. The experiment of Schwann applied to this sugar solution is, therefore, of irreproachable exactitude. Schwann, however, did not always succeed so well as he wished, and the experience of Mantegazza and Pouchet was at variance with his general conclusions ; even Pasteur himself in some experiments failed to preserve his liquids. These are the particular instances :—Five flasks of 250 cubic centimètres capacity, containing 80 cubic centimètres of the sugar solution, were boiled, and during ebullition sealed up. The points were broken under mercury, and pure gases in all cases but one let into the flasks. Organisms were found in every case after four days. In all these experiments, as in those likewise of Schwann, which were contrary to the result of his first experiment with extract of meat, it was the mercury that introduced the germs. In making such experiments with a mercury trough, preservation of the liquid will not always succeed, even if it succeeds sometimes. If the sugar solution be replaced by milk and treated by either of the methods above described, the milk putrefies. These results, so different and contradictory, find a natural explanation farther on, but so far they are facts of a troublesome nature.

*Germination of the Dust which exists suspended in
the Air, in Liquids suitable to the Development
of the Lowest Organisms.*

The facts ascertained so far are :—

1. That there exist suspended in the air
organised corpuscles exactly like the germs of the
lowest organisms.

2. That sugar solutions with the liquor from
beer yeast, a fluid extremely alterable in ordinary
air, remains unchanged and limpid, without giving
rise to bacteria or fungi, when left in contact with
air previously heated.

The question now arises, how is it possible to
sow an albuminous sugar solution with germs
collected by means of pyroxyline in the manner
already described ?

Taking a flask containing such a sugar solution
kept at 77° to 86° F. for one or two months unchanged, in contact with previously heated air, the
sealed-up end is connected by means of a caoutchouc tube with one part of a T-tube, while another
is in connection with an air-pump, and a third with
a platinum tube heated to redness. Between the
T-tube, however, and the flask is a wide tube containing a very narrow one within it, holding a plug
of gun-cotton, through which a large volume of air
has been passed. The tap in connection with the

heated platinum tube was closed, and the one in
connection with the air pump opened : after exhausting, air was admitted through the red-hot
platinum, the tap was closed, and the air again
pumped out, fresh air being admitted through the
heated tube ; this was repeated three or four times.
The stop-cocks were then closed, and the sealed
beak of the flask was broken within the india-
rubber connection ; the plug of gun-cotton was
shaken into the liquid, after which the flask was
sealed up again. All experiments so performed
resulted in the liquid, after three or four days'
exposure to a temperature of 77° to 86° F., decomposing, and being found to contain bacteria,
vibriones, and fungi, exactly like those in flasks
exposed to ordinary air. There was no difference
in the length of time requisite for the change, the
forms of life occurring, or the nature of the change
resulting in flasks so treated, and those with the
same liquid exposed to common atmospheric air.
These experiments can scarcely be surpassed for
beauty in their arrangement, or for the importance
and clearness of the evidence they afford. Yet
thinking that it might be objected that the gun-
cotton had given rise to the changes produced,
Pasteur made use of plugs of asbestos, and found a
like result ; but when the plugs of asbestos were
heated red-hot previous to being put into the flasks

the liquids remained unchanged in every case, and so constantly and with such perfect exactitude after an immense number of trials did the results remain the same, that the experimenter himself was astonished.

Extension of previous results to other very alterable Liquids—Urine, Milk, and Albuminous Sugar Solution mixed with Carbonate of Lime.

The facility with which urine exposed to the air becomes altered, and the change which takes place is well known.   It becomes turbid and alkaline, sometimes filled with bacteria, or covered with patches of *Penicillium glaucum, Aspergillus glaucus,* or *Mucor mucedo.*   Often there is formed, when the temperature is not higher than 59° F., a pellicle consisting of a remarkable mould closely resembling *torula*, but which is believed by Pasteur to be a different species.   It consists of transparent cells, often without a nucleus, and considerably smaller than the cells of beer-yeast.   There is also present in urine, when rendered alkaline by the urea being changed into carbonate of ammonia, a peculiar fungus in necklace-like groups, and this organism Pasteur is fully persuaded is the cause of urea being converted into carbonate of ammonia. An interesting observation was made with regard

to the turbidity of liquids, which generally is the
first sign of alteration ; this is caused not merely by
the presence of minute organisms, such as bacteria,
but by their movements in the liquid ; for when
they are dead they settle to the bottom of the
vessel, and the liquid becomes clear again. Many
flasks of urine were treated in the manner already
described—that is to say, they were boiled, and
heated air was admitted to them. After preservation for months at 77° to 86° F., without change,
plugs of asbestos through which air had been
drawn were introduced ; and then in cases where
the liquid was alkaline, strings of this peculiar
fungus were found invariably, and crystals of
ammonio-magnesian phosphate were deposited.
It was observed that *Bacteria* (see fig. 60) appear
in a liquid before any other organism. These
organisms are so small that it
would be impossible to distinguish their germ ; but even if the
appearance of the germs were
known, it would be still less possible to recognise them among
the various particles of organised dust collected
from suspension in the atmosphere.

Fig. 60. *Bacteria.*
As seen by Pasteur.

In experimenting with milk boiled in flasks and
exposed to heated air, it was found that generally
in from eight to ten days, but in one case after so

M

long a time as a month, the milk was found to be
curdled.   Microscopic examination showed that
the whey was filled with vibriones, often of the
species *Vibrio lincola* (see fig. 59, p. 153) and bacteria.   The air of the flasks showed that the
oxygen was replaced by carbonic acid ; yet swarms
of these vibriones were living in an atmosphere
without oxygen.   The most important observation
which leads to an explanation of the extraordinary
behaviour of milk in these experiments, is the fact
that no mucor, torula, or penicillium—nothing but
bacteria or vibriones—were found in the liquid.
The obvious conclusion is, that these organisms
or their germs are not destroyed by a temperature
of 212° F., when the heated liquid which serves to
develop them enjoys certain properties.   To test
this supposition, the milk was boiled under pressure
so that the temperature was raised during ebullition to 230° F., and then heated air was admitted
of course at the usual atmospheric pressure ; flasks
treated in this way were kept an indefinite period
without the production of any life whatever.   The
milk preserved its flavour, its odour, and all its
properties.   Sometimes a slight oxidation of fatty
matter took place, as could only be expected in
such a considerable body of air ; this was proved by
an analysis of the air.   In such cases the milk had
a slightly suety taste.   But what condition prevents

the development of vibriones in sugar solutions and urine when heated to 212° F. ? It is the fact that they contain a trace of acid. Milk is an alkaline liquid.[1] If a liquid of the following composition :—

Sugar . . . . . · .	10 grms.
Yeast water . . . · .	100 c.c.
Carbonate of lime	1 grm.

be boiled in flasks at 100° C., filled with heated air and sealed up and left to itself at 25° to 30°, in from two to four days it becomes turbid from vibriones, which have a very lively motion. It was found that a species of mucor after a time covered the surface of the liquid. It seems, therefore, that under these particular conditions the germs of this cryptogam had resisted the temperature of boiling water. An important confirmation of these experiments regarding the failure of a temperature of 212° F. to destroy certain germs here follows. Milk which had been preserved some months had a plug of asbestos presumably containing germs introduced into it by the method already described ; it was sealed up, and the flask was then plunged into boiling water ; in eight days bacteria and vibriones were found in swarms. It was further discovered that 226° F. was too low a temperature to insure the preservation of these liquids.

[1] More strictly speaking, milk is one of the few liquids, according to Prof. Lister, which show both an alkaline and acid reaction to litmus.

It cannot be too forcibly impressed on the reader by what means and with what certainty Pasteur demonstrated the fact of myriads of organisms occupying comparatively small volumes of air. This is a point to which his detractors have willingly made themselves blind ; they tell us the organisms are few in number without any experimental proof ; while, on the other hand, as will be presently seen, Dr. Angus Smith and Mr. Dancer estimated that there were $37\frac{1}{2}$ millions of organisms, many of which were recognisable, in 2,500 litres of Manchester air.[1]

Another method for showing that all the Organisms produced by previously heated Infusions have for their origin the particles which exist suspended in ordinary Atmospheric Air.

Says Pasteur, 'I believe it to be rigorously established in the preceding chapters that all the organised productions of infusions previously heated, have no other origin than the solid particles which are always carried in the air and left deposited constantly upon everything. Could there still remain the least doubt of this in the mind of the reader, it will be dissipated by the experiments I will now describe.'

The experiments consisted in placing in glass

[1] 'Air and Rain,' p. 305.

flasks the following liquids, all of which are very changeable in contact with ordinary air : yeast liquor, sugar solution and yeast liquor, urine, beet-root juice and infusion of pears ; the flasks were then drawn out so as to have a long neck with many bends in all directions. The liquid is boiled for some minutes, while the steam escapes plentifully from the open neck. Such flasks may then be left to themselves without being sealed, and, strange to say, though the air enters, the liquid may be pre-served for an indefinite period—an interesting fact for those who are accustomed to make experiments of such a delicate nature as this subject requires. There is no risk in transporting these flasks from place to place, or submitting them to the varying temperature of the seasons ; the liquids show not the slightest alteration in taste or smell ; they are truly specimens of Appert's food-preserving pro-cess. In some cases there was a direct oxidation of the matter, a purely chemical process. But it has already been shown how this action of oxygen was *always limited when organised productions were developed in liquids*. The explanation of these new facts is, that the air on first entering comes in con-tact with water vapour at the temperature of 212° F., and is so rendered harmless ; what follows enters but slowly, and leaves its germs or particles of active matter in the moist curvatures of the tube-

neck. After remaining many months in a warm
place, the necks of the flasks are cracked off by a
file mark without other disturbance, and in twenty-
four to thirty-six or forty-eight hours, fungi and
other organisms make their appearance in the
usual manner.

The same experiments can be made with milk,
but then the milk must be boiled under pressure ;
milk has been kept for months in these open flasks
without change at a temperature of 77° to 86° F.
The production of organisms can always be started
in these flasks by briskly shaking the liquid or by
sealing during ebullition, and after cooling allowing
the air to enter suddenly by breaking the point of
the tube.

Many such flasks, exhibited at the Academy of

Fig. 61.   Solutions in flasks similar to those of M. Pasteur.

Sciences, were preserved with their contents unchanged for eighteen months, although extremely
prone to decomposition.   Here, in the annexed

wood-cut (fig. 61), are represented two flasks similar
to those of Pasteur, and two plugged with cotton
wool after the plan of Schrœder and Dusch ; in the
centre is an open flask of liquid made at the same
time and treated in exactly the same way as the
others, but instead of being still bright and clear like
them, it has now become a mass of brown filth.
When the photograph for this illustration was taken,
the flasks with their contents were three years old.

'The great interest of this method is, that it
unquestionably proves that the origin of life in infusions which have been boiled is solely due to
solid particles suspended in the air.   Neither a gas,
divers fluids, electricity, magnetism, ozone, things
known or hidden causes, there is absolutely nothing
in ordinary atmospheric air which, failing these
solid particles, can be the cause of the putrefaction
or fermentation of the liquids which we have
studied.'   It has so far been definitely proved by
Pasteur, and stated in the following manner :—

' 1st. That there are constantly, in ordinary air,
organised particles which cannot be distinguished
from the true germs of the organisms found in
infusions.

' 2nd. When these particles and the amorphous
débris associated with them are sown in liquids,
which have been previously boiled and which remained unchanged in air previously heated, there

appear in these liquids exactly the same forms of
life as arise in them when they are exposed to the
open air.'

Such being the case, could a partisan of spontaneous generation wish to uphold his principles
even in the face of this double proposition ? He
might, but then his argument would necessarily be
of the following kind, of which I leave the reader
to judge for himself. There are in the air, he might
say, solid particles, such as carbonate of lime,
silica, soot, fibres of linen, wool, and cotton, starch
granules . . . . and besides these organised corpuscles having a perfect resemblance to the spores
of the Mucedines or the germs of Infusoria.   I
prefer to attribute the origin of Mucedines and
Infusoria to the first amorphous substances rather
than to the second.' [1]

This has actually been urged.   Could there
be more eccentric reasoning ?

---

[1] As to the mention here made of *Infusoria*, it must be recollected
that Pouchet stoutly maintained that the true *Infusoria* were subject
to heterogenic reproduction.

It is not exactly true that the smallest quantity of ordinary Air gives rise in an Infusion to the Organisms peculiar to this Infusion. Experiments on the Air of various Localities. Inconvenience of employing Mercury in Experiments relative to Spontaneous Generation.

If the smallest quantity of air in contact with an infusion gives rise to organisms, and these organisms are not of spontaneous origin, then it follows that in the minute portion of air there must exist a multitude of the germs of very different organisms ; in such numbers, too, that, as Pouchet says, the air would be so loaded with organic matter as to form a thick fog.   Strong as this reasoning is, it would be still stronger if it were shown that different forms of life are derived from different germs : this may be so, but it has not been proved.

Experimental proof of this statement, the error in which lies in gross exaggeration, was made by sealing up during ebullition flasks of 250 cubic centimètres capacity containing about 80 cubic centimètres of various liquids.   On breaking the points of these flasks in certain noted places, the air entered with a rush into the empty space, carrying the germs along with it ; after resealing, the flasks were placed in a warm situation and any change noted. In some cases the decomposition followed, and the production of the usual forms of life ; in other cases the flasks remained as if they had been filled with heated air, quite unchanged. In two experiments made in the open air after a slight shower in the month of June, both resulted in the production of organisms ; in four others, after a heavy rain in the same place, two of the flasks preserved their contents unchanged for at least thirteen months afterwards. These experiments were made, it is easily seen, in an agitated air, but Pasteur carried his labours into the cellars of the Paris Observatory, where the air is quite still except when agitated by the movements of the experimenter, and in that region below the surface of the earth where the temperature is unaffected by the changes of the seasons. It is to be expected that air, in which there is so little to cause its disturbance, would have deposited on the ground the germs which at one time floated in it. A greater proportion of flasks, therefore, if opened and resealed in such an atmosphere, should have their contents preserved. Out of ten experiments made under such conditions with yeast water, in only one was any living thing found ; while eleven experiments made in the court-yard of the Observatory at a distance of 50 centimètres from the ground, and at the same time, rendered in every case the usual forms of life ; a modification of these trials was made by letting air into flasks of liquid at various mountain heights.   Eighty-three flasks, prepared in the manner already mentioned, were experimented on : twenty of these were filled up with air at the foot of the heights which form the first plateau of the Jura ; twenty others on one of the peaks of the Jura, 850 mètres above the sea-level ; and the remaining twenty were carried to Montanvert, near the Mer de Glace, at an elevation of 2,000 mètres.   The result was, that of the twenty opened on the lowest level, eight contained organisms ; of the twenty on the Jura, five only contained any ; and lastly, of the twenty filled at Montanvert, while a strong wind blew from the deepest gorges of the Glacier de Bois, one only was altered.   The method of opening the flasks was to hold them with outstretched arms and the point of glass turned from the wind, and by a pair of iron forceps, which had just been heated in a spirit-lamp flame, the point was broken.   The drawn-out point had been previously scratched with a file and heated ; otherwise, particles of dust adhering to the glass would have been carried into the liquid by the in-rush of air.

A remarkable and interesting fact connected with these experiments was, that on one occasion Pasteur opened his flasks, and, on account of not being able to see the flame of his lamp against the brilliancy of the snow, it was impossible to re-seal them ; the flasks were necessarily carried back to the little inn at Montanvert to be closed up. Every one of these flasks contained organisms after keeping for a short time. On the Glacier, then, there are no germs in the atmosphere, but at the neighbouring inn the air swarms with life, and life from all parts of the world, brought by the travellers.

Explanation of the Cause of Failure of the Experiments in which Mercury is used.

Flasks containing liquids which had been kept for a great length of time were connected with an air-pump and a red-hot platinum tube : after repeated exhaustion and re-filling with heated air, the communication was made between the flask and the platinum tube, and a globule of mercury taken out of a mercury trough in a laboratory, which had previously been introduced into the connecting-tube of india-rubber, was made to roll into the flask ; on re-sealing and keeping for a few days, fermentation ensued in every case, just as certainly as when the asbestos plugs and the adhering germs were sown in similar liquids. This case leaves no doubt regarding the cause of failure of experiments in which the liquid comes in contact with mercury by the flasks being broken under the surface of the quicksilver.

There are other facts which Pasteur established, of great interest and importance in connection with the nutrition of minute fungi. Instead of experimenting on milk, urine, or solutions containing the liquor from yeast, he made use of such an infusion as the following; that is to say, a mixture of perfectly definite chemical substances :—

Pure water	100 parts
Sugar-candy	10 ,,
Tartrate of ammonia . . .	0·2 to 0·5 part
Ashes of yeast	0·1 part

On impregnating such a liquid, when supplied with heated air, with germs collected from the atmosphere, bacteria, vibriones, and fungi, &c., were soon developed; the albuminoid and fatty matters, the essential oils, and pigments belonging to these organisms being derived from the elements of the ammonia salt, the phosphates, and the sugar. These complete organisms were built up out of the material afforded by such a mixture of simple substances, a fact which is quite contrary to Pouchet's declaration that ovules or germs were evolved from a sort of vitality remaining in lifeless, or, rather, dead matter—that is to say, matter deprived of life.

A solution consisting of—

Pure water	100 parts
Sugar-candy	10 ,,
Tartrate of ammonia . . .	0·2 to 0·5 part
Yeast ashes	0·1 part
Pure calcium carbonate . .	3 to 5 parts

showed much the same phenomenon, in fact, differing only by a more marked tendency towards
the changes called lactic, viscous, and butyric fermentations ; and all ferments, whether vegetable
or animal, characteristic of these changes, were
produced, simultaneously or successively.

Professor Tyndall, in 1870, gave us a means of
investigation, supplementary to the microscope,
and of extreme delicacy. Aided by Professor
Huxley, he proved that particles in a liquid, quite
invisible under an object-glass readily showing
bodies $\frac{1}{100000}$ of an inch in diameter, were revealed
with the greatest ease by means of a beam of light.
If the air were pure, a beam of sunlight traversing
a darkened room would be invisible except where
it struck upon the wall. It is the scattering of the
light by the floating dust which makes the track
luminous ; the larger and more numerous the particles the greater the luminosity. Hydrogen, coal-
gas, air passed through cotton-wool, and the air of
still places, were found to be free from floating
matter. Having devoted much attention to this

subject since 1865, I made use of this discovery to aid me in a very careful repetition of some experiments published by Dr. Bastian.

The following description of the method pursued in this research is taken from the ' Proceedings of the Royal Society for 1872 : '—

Experiments concerning the Evolution of Life from Lifeless Matter.

On June 30, 1870, there appeared in ' Nature ' a paper by Dr. Bastian, entitled ' Facts and Reasonings concerning the Heterogeneous Evolution of Living Things ; ' the perusal of this, and its continuation, led to the belief that another interpretation might be put on the results obtained by Schwann, Pasteur, and others, not so much by virtue of the arguments made use of, as by accounts of experiments given in detail.   The most remarkable case was that of Exp. 19, in which the author gave a drawing of a large organised mass obtained from a solution of sodium phosphate and ammonia tartrate, which had been exposed to a temperature varying between 294° F. and 307° F. for four hours. This organism was seen to grow within the flask till it attained a certain size, beyond which it did not increase.   Now a fact so distinctly stated as the production of an organism, and its development to a considerable size, from a liquid containing nothing further than phosphate of soda and tartrate of ammonia, in a flask from which the air had been most thoroughly withdrawn, and which, when containing the liquid and hermetically sealed, had been heated to so high a temperature, was (admitting the conditions and performance of the experiments to be faultless) an absolute proof of the evolution of living matter *de novo*.

While adhering as strictly as possible to the conditions of Dr. Bastian's experiments, it was necessary to devise some refinement on the mode of examining the liquids experimented on without exposure to atmospheric air, for even the presence of one single fungus spore might be regarded as evidence of this process of generation.

The most promising plan seemed to be to open the sealed vessels in an atmosphere artificially prepared so as to be free of living matter. Hydrogen being fourteen times lighter than common air, may remain in contact with it without risk of contamination by floating matter ; indeed, Professor Tyndall's demonstration, by means of a powerful beam of light, that such an atmosphere is free from dust, was sufficient to warrant its use. The means whereby this fact was made of further practical value are the following :—

1st. *The experimental tubes* in which the infu-

sions and solutions were heated were made of ordinary combustion tubing drawn out at the lower
end, first to a finer tube ¼ the diameter of the
original, and after a space of an inch or so to
extreme smallness. The solution or infusion was
then made in a flask with distilled water, drawn by
a siphon from a carboy after standing at rest for
many hours, the syphon dipping into the middle of
the liquid. The flask was, after the usually careful
cleansing that chemical vessels require, rinsed out
with a solution of potassium dichromate, mixed
with strong sulphuric acid, then washed out with
distilled water. A further quantity of distilled
water collected in this vessel was used for the
solution. The experimental tube, with the lower
end drawn out but open, was cleansed with acid
dichromate, and afterwards with hot distilled water.
The fine point was open, in order to let the liquids
run through, otherwise a drop might have collected
in its capillary portion, and, not being dislodged,
would have interfered with the experiment. This
tube, then, best described by fig. 1, p. 178, had the
fine end sealed up, and the liquid to be experimented
on poured in till it about half filled the tube. The
upper part was then drawn out, so as to serve a
purpose yet to be named, and also in such a manner
that it could be adapted to the Sprengel exhauster ;
after this operation it had the form of fig. 2. It was

then fitted to the Sprengel exhauster by means of a tube shown in fig. 3, which admitted of connection by means of two Sprengel joints, the one over the experimental tube being made air-tight with water. The reason for water being used was this : had there been a leak, water would have entered the tube, and so no damage would have resulted ; but

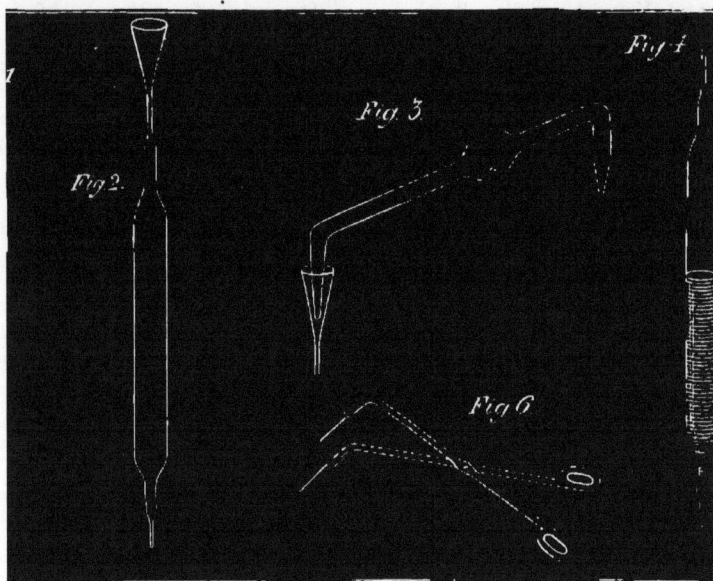

Fig. 62.

had it been mercury or glycerine, the tube would necessarily have been rejected.   The use of the bulb on the connecting-piece was to catch the water which boiled or distilled over during the exhausting process. After complete exhaustion, recognised by the clicking of the falling mercury, a blowpipe-flame was cautiously applied to the fine tube till it fused thoroughly, and it was then drawn away from the other portion. What was so far accomplished was the sealing of a solution or infusion in a vacuous tube of the annexed form (fig. 4). The tubes thus prepared were heated in an air-bath, consisting of horizontal iron pipes surrounded by two iron jackets. The tubes lying horizontally were not in contact with the liquid on the upper part ; so, after being heated in the one direction for a time, at Professor Tyndall's suggestion they were turned over and heated anew, so as to bring every part of the tube into contact with the heated liquid. These tubes had generally a bead of glass fused to one side, so that if the tube were heated with the bead uppermost, it had to be turned over and heated again, the bead being downwards. Generally four tubes were heated at a time ; and one of these was soon after cooling opened to allow access of air, in order to observe whether any change occurred differing from any that might take place in the sealed tubes.

2nd. *Apparatus for Examining the Contents Microscopically out of Contact with Air.*—This consisted of a bell-jar closed at the top with a bung, and supported on a tripod ; this bell-jar was kept filled with hydrogen by means of a self-regulating apparatus continually passing a gentle stream of gas into the upper part of the jar by means of a glass tube. The bung was bored with an eccentric hole ; through this passed the tube containing the liquid for examination ; the end of the tube passed very little below half-way down the bell-jar.

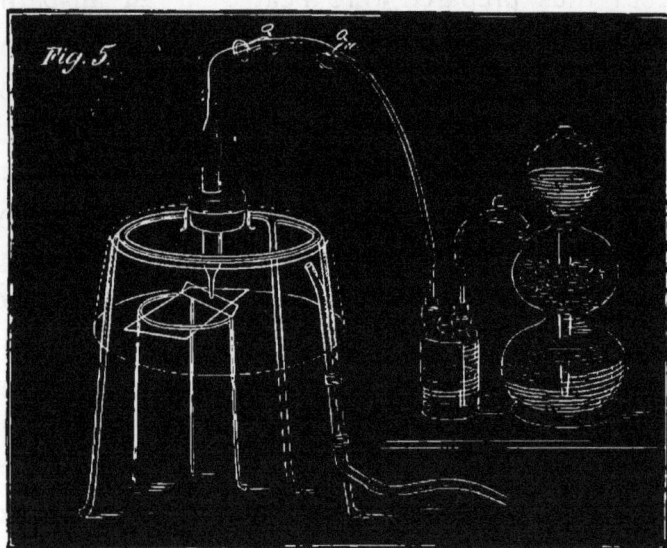

Fig. 63.

Under this was a small tripod, on which rested a glass plate to be used as a stage for the microscope slips. The tube then being *in situ*, over the upper point was tightly slipped a piece of non-vulcanised india-rubber tube, connected with a constant hydrogen generator. It was not deemed advisable to make use of coal-gas, because, had any lifeless

organism been found in a tube, it might have been
objected that a trace of benzole or naphthalene
vapour or other impurity had been fatal to the
experiment. On the india-rubber tube were two
burette clips; now, by breaking the fine point
within the india-rubber (a scratch with a file being
previously made upon it), hydrogen flows into the
vacuous tube. Both clips are now closed, and by
means of forceps, the hand being beneath the bell-
jar the lower end of the tube is broken off. No
liquid, however, escapes, because the internal pressure is not much in excess of that of the atmosphere.
The condition of things now can be explained only
by the aid of a drawing (fig. 5—in fig. 63—p. 180 ;
the whole arrangement consists of a pipette containing the experimental liquid, above and below
which is an atmosphere of hydrogen ; each drop of
liquid expelled is received on a slip of glass in such
an atmosphere. A drop of liquid is deposited by
squeezing the space of india-rubber between the
two clips, that clip nearest the bell-jar being open :
before removing the pressure, it is again closed,
and the partial vacuum made by compressing the
india-rubber is filled up by allowing gas to flow in
from the apparatus. This precaution prevents the
possibility of the atmosphere of the bell-jar bub-
bling up into the tube after a little fluid has been
discharged.

3rd. *Treatment of the Glass Slips.*—These are heated in an air-bath to about 390° F., taken out while hot with forceps, and placed on the stage in the hydrogen to cool, and kept there till wanted.

4th. *The Glass Covers.*—These are washed in the acid-dichromate solution, then in distilled water, and finally in alcohol, and being picked up by a pair of peculiar forceps, are heated over an argand burner till very hot ; they are then held in the glass vessel full of gas till cool enough to use. The forceps I devised (fig. 6—in fig. 62—p. 178) have points of watch-spring steel, so that a thin glass cover may be firmly gripped without breakage. They are made by cutting a small pair of crucible tongs an inch or so before the part where they bend at right angles ; they have then two pieces of thin brass riveted on, which are bent at right angles an inch or so from their ends ; the points are made by riveting on pieces of watch-spring steel a little more than an inch in length. When these tongs are held in the position of scissors, that is to say, with the thumb above the forefinger, the ends point downwards. To take up a glass cover, the forceps being in the position mentioned, the wrist is turned over from right to left while the elbow is raised, the glass resting on the lower point while the upper is closed down on it, may be safely held and conveyed to where it is wanted.   After a little practice these pincers are easily used.

The advantage of the bung closing the neck of the bell-jar being bored eccentrically is, that by simply turning the bell-jar horizontally the pipette point delivering the liquid may be shifted from a glass slip on which a drop of liquid has been deposited to another clean one, or be made to turn through such an angle as to be out of the way of the glass stage, in order to transfer the solution to a flask for further experiment.

Tartrate of ammonia was prepared by neutralising a tartaric-acid solution with aqueous ammonia; this was mixed with phosphate of soda solution, made by dissolving carefully washed crystals of the salt in hot distilled water.   The mixture containing about 5 per cent.[1] of the two salts was slightly acidified with tartaric acid.   It was found that in no case should filtering be resorted to if possible, as the finest Swedish paper transfers myriads of its fibres to the liquid.   With proper care, however, filtration is unnecessary when dealing with solutions of definite salts.

---

[1] No mention was made by Dr. Bastian of the strength of his solutions, so that this, it is natural to suppose, he considered of no importance.   In order to guard against error, these liquids were always exposed to the air to ascertain the fact that fungi would be developed in them, and some form of life was found to make its appearance in every case.

Modifications of Experiments.

In the renewed examination of liquids kept some time in sealed tubes, commenced in July 1871, a slight modification in the original method of proceeding was used. A bell-jar was chosen, the upper mouth of which was ground perfectly flat at the edge. Instead of inserting a bung with a hole in it to receive the tubes, a metal disk, with a wide metal tube placed eccentrically and projecting half an inch, was luted on to the mouth of the jar by means of grease, or, better still, what is known in pharmacy as *resinæ ceratum*. The glass sealed tube was then slipped into an india-rubber conical stopper, or rather ring;[1] for the thickness of it was so slight that a tube of any size could be made to fit it, either by the india-rubber stretching when the tube happened to be large, or by binding with a piece of copper wire when it fitted loosely. The pipette was scratched with a file at each end, and over the upper one was slipped a piece of india-rubber tube, attached to a tube of glass about $\frac{1}{4}$ inch in bore and 4 inches long, tightly packed with cotton wool. The caoutchouc tube was pinched by a burette clip, and the extremity of the tube enclosed by the caoutchouc was broken

---

[1] These things are made and sold for the purpose of fixing the taps into beer-barrels.

at the file-mark. The vacuum was considered good if the india-rubber tube collapsed completely ; the burette-clip was opened, and filtered air thus admitted into the vacuous space. In order to render anything that might be attached to the interior of the india-rubber tube harmless to the experiment, it was dipped in glycerine and the glycerine squeezed out of it, or treated in the same way with melted bees'-wax or paraffin. The pipette fits into its place in the disk by means of the flexible stopper. By closing the burette-clip, the tube can be broken at the lower point without more than a drop or two of the liquid escaping. After about one-third of the liquid had been examined, one-half of the remainder was allowed to run into a flask which had been previously heated to between 390° F. and 570° F.[1] The tube was then removed, and the fine capillary point, when possible, sealed at a gas flame. The finer the point the more easily is this accomplished. A portion of liquid remains in the tube. On heating rather strongly a little of this is driven out, and then no air can pass to the remaining liquid without passing over red-hot glass, which readily melts together. The tube and flask were then placed side by side in a warm place to undergo further

---

[1] That is to say, baked in an oven the bottom of which was red-hot.

observation. If the tube, or class of tubes, were called A, after opening it was labelled A′, and the liquid out of it exposed to the unfiltered air, A″. The tubes and flasks labelled thus were kept in a cupboard, the bottom of which was the metal lid of a long water-bath. It was thought better not to place the flasks or tubes in water, because the aqueous vapour which would thus surround the mouths of the flasks would create an abnormal atmosphere which might or might not affect the experiments ; besides, such a plan is not so cleanly. The objective made use of was obtained from Messrs. R. and J. Beck. It was a ⅛ glass, without any immersion arrangement, and gave, with the second eyepiece of one of their miscroscopes, a magnifying power of 750 diameters. Occasionally, for convenience in drawing, a power of 420 diameters was employed.

Method of examining a Liquid which it was difficult to retain in the Pipette-tube.

When it happened that the finely drawn-out end of the pipette was too large to retain the liquid, it was allowed to run into a small glass vessel, really a beaker cut down so as to measure about 1½ inch in diameter and 1 inch high. Drops of the solution were removed from this to the glass

slides while it stood on the glass stage in the jar of
hydrogen, by means of a tube like a very long-
legged syphon, at the lower end of which was a
piece of caoutchouc tube, with one end stopped by
a little piece of glass rod.   This was nothing more
than a bent pipette ; by compressing the india-
rubber when the pointed tip of the shorter limb
was dipped into the beaker-glass, and then releasing
it, the liquid entered for the space of an inch or so,
and could then be easily transferred to a glass slip.
It was thought as well to blow hydrogen through
the tube before use ; and, of course, like all the
other apparatus, it was carefully washed and
heated.

Preparation of Solutions.

The water used was very pure distilled water
taken from a carboy, the contents of which had
been tested with a beam of light, and found to
reflect chiefly the blue rays. A previous attempt
to obtain pure water by distillation with sulphuric
acid and potassium permanganate, in glass vessels
and an atmosphere of hydrogen, did not yield
better specimens. It is impossible to prepare solutions of salts which do not show abundance of
floating matter to a ray of light, even when such
pure water is made use of. Solutions filtered
through the finest Swedish paper are crowded with

fibres, which may readily be seen by filling a globular flask with the solution, the eye and an argand burner being on the same horizontal line, and about a foot apart. The flask is interposed, and gradually lowered till the particles are seen brilliantly illuminated on a dark ground. The phosphate of soda used was recrystallised immediately before being dissolved, and the tartrate of ammonia was prepared from recrystallised tartaric acid and the strongest aqueous ammonia. When the solutions were mixed, the alkaline reaction was neutralised by tartaric acid, or rendered faintly acid. Of course it is of the first importance that the tubes, after being sealed, should be heated immediately to the temperature necessary to destroy life, and this was done in every case as soon as possible.

Without entering into details concerning each particular tube, it may be stated shortly that in the liquids when kept in vacuo for from six to twelve months, and afterwards exposed to filtered air for periods of from two months to a year, nothing at all was discovered, while those portions of the liquids let out into perfectly clean flasks, previously heated to the temperature of boiling mercury, became always filled with life.

The figs. 64, 65, and 66 were noticed in the contents of different tubes; also fig. 58 B, p. 152.

It may doubtless appear to some that this
particular mode of experimenting involves the

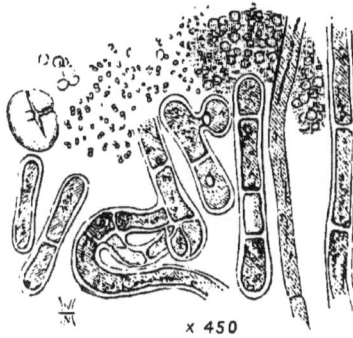
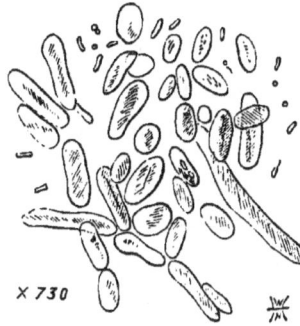

Fig. 64. Confervoid Growth.   Fig. 65. Torula (rough sketch).

introduction of needless complications, that others
have obtained the same results without the use of
such apparatus ; but the incorrectness of this will
be admitted when it is pointed out that the necessity of being guarded, of being safe, indeed, from

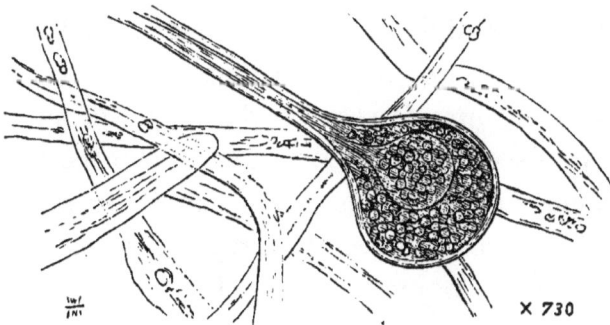

Fig. 66. A species of Mucor.

every possibility of error, is absolute.   Had living
things been discovered, that the experimental

method and apparatus was a matter of the first importance would have been evident, inasmuch as it excluded all possibility of atmospheric contamination of the experimental liquids during examination. In a word, it was necessary to be prepared for any result, and be guarded on every side.

The appearances described in Exp. 20,[1] by Dr. Bastian, are, with the exception of the fungus-spores and bicellular bodies, exactly what one sees in silica. Having ascertained the fact that phosphate of soda, and especially when not neutralised with acid, attacks glass tubes at a temperature of 302° F., and as in this case sodic phosphate and ammonic carbonate [2] were heated for four hours at 294° F. to 307° F., some of the silica deposited from these experimental tubes was examined, and gelatinous matter, resembling that met with in solutions containing organisms, was noticed : there were two or three transparent spherules also, most probably drops of water enclosed in silica ; they are often seen in pectised silica. As for the matter becoming stained by magenta, that is no evidence of any organic nature, this property being shared by silica.[3] It may further be remarked that magenta would be precipitated on addition to such

1 'Nature,' vol. ii. p. 200.

2 The reaction of this liquid is described as being neutral, which is evidently a mistake, because both these salts are very alkaline.

3 'Journal of the Chem. Soc.' vol. ix. p. 452.

a solution by the alkaline phosphate and carbonate ;
those parts more deeply stained than the others
would be those where rosaniline was precipitated ;
it would be impossible to use a salt of rosaniline
for the purpose of detecting albuminous matter.

Dr. Bastian considers he has established by
experiment the theory that living organisms,
amongst which are vibriones, and fungi of the
genera *Mucor*, *Penicillium*, and *Torula*, and algæ,
such as conferva, are evolved *de novo* from lifeless
matter ; he brings together a number of reasons,
of a more or less decided kind, to show not only
why he thinks it should be an intelligible process,
but also why others, and particularly M. Pasteur,
have obtained results leading directly to an opposite conclusion.   These arguments are not drawn
from experimental evidence ; they do not therefore
fall within the bounds of this discussion ; but it
should be observed in one case not only do Dr.
Bastian's own experiments deny the truth of a most
important assumption of the evolutionists, but also
at the same time these experiments prove the
contrary.   He says : [1]—'The disruptive agency of
heat is fairly enough supposed by the evolutionists
to destroy some of the more mobile combinations
in each solution—to break up more or less completely, in fact, those very complex organic pro-

[1] 'Nature,' vol. i. p. 176.

ducts whose molecular instability is looked upon as one of the conditions essential to the evolutional changes which are supposed to take place.' Before granting such a supposition, it would be necessary to know, first, what are 'the very complex organic products' of such peculiar 'molecular instability' existing in a solution of tartrate of ammonia, sodic phosphate, acetate of ammonia, oxalate of ammonia, in a solution of sugar and calcined yeast, in turnip infusion, or any other putrescible liquid. These experiments show that there is no such disruptive agency in a high temperature, that it does not influence the 'more mobile combinations,' either in solutions of organic salts or vegetable infusions; for certain tubes in which the ordinary forms of mould developed were the same as the others of the series, contents identical, heated at the same time to the same temperature, in fact taken from among them indiscriminately, the only difference being that one was exposed to the air and the others were not. Besides, there is, in addition, the evidence afforded by the tubes being kept, firstly, in a vacuous condition, secondly, supplied with filtered air, and, finally, freely opened ; yet we find the changes occurred in them as readily as in the unheated original solutions. *Dr. Bastian records*[1] *the development of organisms in a liquid*

[1] 'Nature,' vol. ii. p. 200.

*heated as high as 307° F.; yet the assumed 'disrup-
tive agency of heat' is supposed to have influenced the
results of Schwann and Pasteur at a temperature of
212° F.!* His experience is contradictory to his
own theory, and at the same time to the experiments of others, to which his theory raises objection.

If it be asserted that dead nitrogenous organic
particles in the air, and not germs, are the cause of
organic changes, how are we to account for the
action of various antiseptics? In the Report to
the Cattle-Plague Commissioners 'On Disinfection
and Disinfectants,' by Dr. Angus Smith, an account
is given (p. 10) of the action of a number of essential oils, such as oil of bitter almonds, oil of mus-
tard; amylic alcohol, cresylic and carbolic acids,
and ether, the vapour of which was diffused in air
surrounding pieces of meat. Many substances had
the property of preserving the meat for a great
length of time, especially amylic alcohol and oil of
bitter almonds; chloroform and carbon tetrachlo-
ride also share this property. If these nitrogenous
particles are not living things, how are we to account
for the action of these substances? It cannot
be a chemical action, because the substances are
chemically inactive. In some cases the less active
preservative agents having diffused away out of the
bottle, mould formed on the meat, or putrefaction
commenced; and this always happened on that

part nearest the cork, showing that particles causing the change came in with air after the preservative agent had escaped. That these preservatives
were fatal to the growth of mould, or the spread of
putrefaction, was ascertained by placing mouldy
paste under a bell-glass the atmosphere of which
contained a small quantity of the vapour.

If those particles which are the origin of life
are themselves lifeless, they could not influence the
nature of the organisms developed ; in that case
different solutions would give rise to different
organisms, and, conversely, the same solution
would yield the same form of life. But we do not
find this to be the case ; for in the same infusion of
turnip, there are found on exposure to the air, even
at the same times, different forms of life. In the
one case minute vibriones and masses of *torula*-cells
occurred, whereas in the other confervoid growth
and some minute motionless organisms which had
not previously been observed were the most notice-
able forms. With different portions of a solution
of alkaline phosphates and tartrates, in one case we
got a fungoid and in the other an algoid form.
Were it not for the great preponderance of sound
experimental evidence to the contrary, it would
be easier to believe the theory that life was evolved
de novo. It is the crude idea which a superficial
observer of everyday phenomena would entertain ;
the popular error that maggots are bred out of

corruption illustrates this. The theory involves the discovery of a new property of matter, the property that certain compounds (undefined nitrogenous particles in the atmosphere) must have of decomposing molecules of other substances with which they are in contact, and building out of their constituent atoms, substances of a much more complicated nature, without the exertion of external forces ; even beyond this, they must be capable of arranging those compounds into definite forms. On this account a very great deal of thoroughly sound experimental evidence is necessary to establish the doctrine of evolution of life *de novo*. But, so far as our present knowledge guides us, whether we term it spontaneous generation, abiogenesis, or archebiosis, the process by which living things spring from lifeless matter must be said to be only ideal.

This is so far a *résumé* of the discussion of this subject published in 1872. Since then Dr. Bastian has written two books, ' The Origin of the Lowest Organisms ' and ' The Beginnings of Life,' the direct reasoning of which rests on his own experiments, which, placing the highest possible estimate on them, may be said to be insufficient. Perhaps the further discussion of one or two points may not be without interest.

If a little liquid from one of those tubes which were exhibited at the Royal Society when the foregoing paper was read is let out even now into a clean superheated flask, although at first perfectly clear, sweet, and fresh in flavour and odour, there soon appears a growth of mould upon the surface, and it acquires a very foul smell.   Indeed, these experiments prove, 1stly. That Dr. Bastian probably mistook silica for organic matter in at least one case ; 2ndly. That there is no so-called disruptive agency in a high temperature ; 3rdly.   That hay infusion contains at the outset so many organised corpuscles of various forms that no safe and conclusive experiments on the subject of evolution *de novo* can be made with it ; 4thly. That fungi and bacteria are not generated in infusions previously deprived of all living organisms under any circumstances unless unpurified air gains access.

It is impossible to obtain infusions, and even solutions of crystalline salts free from organic impurities ; still it is possible and indeed necessary to free them from solid particles of any considerable size, and it may here be remarked that Dr. Bastian's flasks being merely washed with hot water could not be clean.   No experiments can be called trustworthy if they have been made with liquids containing solid lumps of dirt or organic matter, such as cheese or muscular fibre.

Cheese, from its greasy nature, would not easily come in contact with the surrounding fluid, and from its slow conductivity of heat, would take a long time before reaching the temperature of the surrounding fluid. The most minute particles might indeed be those least likely to get heated, by reason of air bubbles surrounding them, as is the case with Lycopodium spores and powdered arsenious acid when thrown into water. Yet Dr. Bastian, in his ' Beginnings of Life,' draws conclusions as to the evolution of living things generally from lifeless matter by his own experiments on turnip infusion, containing solid visible lumps of cheese, which mixture had been boiled merely at the ordinary temperature for so short a period as ten minutes.

The immense importance of Pasteur's work can only be appreciated when we consider that he unquestionably established the fact not only that there exists in the air a very large number of *living* germs of divers species—a fact in itself quite sufficient to make us extremely cautious in admitting the theory of heterogenesis—but furthermore that liquids protected in every way against contact with these germs could be preserved for any length of time.

The arguments lately raised that Pasteur used too low a microscopic power in his experiments really do not hold good : instead of weakening they only make his case so much the stronger. The change which takes place in a liquid when it becomes charged with living organisms is perfectly apparent to the naked eye, as well as evident to the nose in most cases. If Pasteur detected so many organisms floating in the air, with a magnifying power of 350 diameters, what must have been the number brought under his notice had he used a glass magnifying 1,000 diameters ? Dr. Bastian used an immersion glass magnifying 600 times (linear), and the particles of matter seen in his liquids were in many cases so trifling that they could very easily have come from the air. Very strong evidence that in some cases this was their origin, exists in the fact that his method of opening his flasks was identical with that employed by Pasteur for infecting his liquids with the germs of different localities, viz., to let air rush into the vacuum.

199

CHAPTER V.

Examination of Arguments raised in Opposition to Pasteur's Conclu-
sions—On Saline Solutions and the Formation of Crystals—The
Effect of Atmospheric Dust on Crystallisation—Crystals and
Organisms—The Antiquated Doctrine of Catalytic Change—Prof.
Tyndall's Recent Experiments—On Arctic Dust—Researches on
Atmospheric Dust by Dr. Angus Smith and Dr. Cunningham —
On the cause of Hay-fever—On the Ageing of Roquefort Cheese—
Nourishment of Plant and Animal Life—Relation of the Minute
Forms of Life to Putrefaction and Decay—Conclusion.

ALTHOUGH it is obvious that the arguments deduced by Dr. Bastian from his investigations can
scarcely receive here any full discussion, more
especially as a careful repetition of his experiments
has failed to afford that corroboration which is
absolutely necessary before reliance can be placed
on them, in short, before they can be quoted *as
evidence*, yet it may be worth while to examine
some of his objections to Pasteur's conclusions and
some of the arguments he considers favourable to
or explanatory of the doctrine of evolution of living
matter *de novo*.

There is one striking characteristic in his method
of reasoning : his premises are all *possibilities*, while

he exacts from others not merely *probabilities* but *absolutely proved facts.*

Thus on pp. 93-95, ' Origin of the Lowest Organisms,' he says, in speaking of the preservation of liquids by means of cotton wool, ' The plug of cotton wool or the narrow and bent tube may, it is true, protect the boiled fluid from subsequent contact with living germs ; but that the liquids do not undergo change on account of such deprivation cannot be safely affirmed, when the same means would also filter off from the fluid some of the multitudinous particles of organic matter[1] (dead) which the air undoubtedly contains, and which may act as ferments.' Which is the more probable, that living or dead matter should act as a ferment ? We know that living matter does, while we know of no dead matter in the air that can do so.

The paragraph above quoted takes it for granted that there are no living germs in the air, ignoring the fact altogether that not only Pasteur proved this, but everyone who has investigated the subject, including M. Pouchet, has confirmed this observation. In short, taking the statements of Bastian and Pouchet together—that there are living germs in the air, and that fermentations take place by means of living organisms—the former wishes us to believe that it is not the living parti-

[1] ' Origin of the Lowest Organisms,' p. 21.

cles which give rise to fermentation, but the dead,
the cotton fibres, the starch grains, bits of wool,
hair, grit, and soot. There is nothing to justify
such a belief. If the dead particles, let us say, for
the sake of argument, initiate a fermentation, they
would do this after the living ones had been killed
by boiling water; consequently the liquids in
Schwann's apparatus would not be preserved. On
p. 98 he seeks refuge from the action of heat in
assuming merely that 'molecular mobility,' whatever that may be, of these 'dead nitrogenous particles is impaired by the agency of heat.' Which
is, again, the most reasonable, that living or dead
matter should thus be changed? There comes on
p. 98 this statement: 'Portions of organic matter
can always be demonstrated amongst such atmospheric dust; whilst living bacteria, or other organisms, such as are first produced by the supposed
sowing of spores, either cannot be demonstrated,
or would seem to be proved by other evidence to be
very sparingly distributed.' Compare this with Pasteur's statement concerning atmospheric dust, also
that of Dr. Cunningham, Mr. Blackley, and Dr.
Angus Smith, which has yet to be referred to,
and then its fallacy is evident. The ambiguous
expression 'organic matter' can only be translated
as germs.

Pasteur says of the *bacterium termo*, 'This

organism is so small that its germ must be indistinguishable, and if its germ were known it would not be possible to recognise it amongst the very various and numerous particles of organic dust collected from the air.'[1] The late researches of Sanderson and Ferrier prove that bacteria germs are invisible under the highest powers of the microscope.

But to return to the examination of Dr. Bastian's arguments ; there occurs the announcement on p. 101, ' Origin of the Lowest Organisms' : ' Some of the very fluids which remain pure in the bent neck apparatus will become fetid if shut up in vacuo.' This is only Dr. Bastian's assertion deduced from his own experiments. My experiments completely disprove this, as I made use of the same liquids as he did. Again, if we grant it, how can the nitrogenous particles in the atmosphere affect a liquid in vacuo? If some liquids, and they include, according to the experiments already quoted, infusion of turnip, and ammonium tartrate with di-sodic phosphate solution, are capable of yielding life without the action of dead nitrogenous organic particles, it is evident that when they are kept at first in vacuo, and subject to only filtered air, no objection can be raised against them whatever, yet they never showed signs of living things until exposed to the open air.

[1] 'Ann. Chim. Phys.' lxiv. p. 57.

With regard to any supposed analogy between the origin of organisms and the origin of crystals, a moment's consideration is sufficient to show how entirely it is without foundation (p. 108), 'living matter develops into organisms of different shapes, whilst crystalline matter grows into crystals of diverse shapes.' This sentence cannot be said to throw any light on the origin of life, for except in the fact that a difference of form results in both cases, there is not the slightest analogy between the cases. When an organism is produced in a liquid there is first of all a new kind of matter or several new kinds of matter formed by the splitting asunder of the compounds already existing, and a re-combination in a different way and in a more complete manner. These newly formed compounds are then moulded and distributed in a certain form, and this form has the power of performing definite functions; it can decompose ready formed compounds, and re-combine their elements in a different way so as to add to its own substance by the processes of growth and nutrition. The development of an organism in a liquid involves very complicated analyses and syntheses. In the crystallisation of a solution nothing similar occurs. We have a solution of a definite substance; this solution is in unstable equilibrium between the solid and the liquid states of aggregation; by a slight disturbing cause of a physical nature, very often such as change of temperature or vibration, the balance is upset and the substance becomes a solid. We know perfectly well that sulphate of copper will not crystallise with 5 molecules of water until its solution has been evaporated down so that it practically has the composition of the crystals *plus* a solution of the crystals in water. So again with nitrate of potassium. When the crystals are formed it is because the whole liquid has the composition of a solution of nitre *plus* nitre, which separates out; or with alum a solution on the point of crystallising consists of a solution of the salt *plus* the actual substance in a liquid condition. When the temperature is lowered the substance no longer retains its liquid state, and in assuming the solid it takes a perfectly definite form. There are no analytical and synthetical processes at work here; the only change is from the one state to the other, from the liquid to the solid. Something analogous occurs when a gas is dissolved in water: it is in a state of unstable liquefaction, and a slight disturbing cause makes it assume its gaseous condition.[1]

[1] Soda-water and champagne afford us examples of unstable solution. They contain carbonic acid dissolved under a high pressure; when the pressure is removed all the gas does not immediately escape. The particles of dust which collect at the bottom of the hollow stem of a champagne glass cause it to throw up a fountain of foam for some time, and a lump of sugar thrown into a glass of nearly still soda-water will renew its effervescence somewhat briskly.

The effect of solid particles on the crystallisation of solutions is strikingly shown by the following experiments derived from the results of Lœwel
and Schrœder and Dusch. A large flask containing a gallon of alum solution is uncovered, and
observation will show that the first trace of dust
which touches the surface causes a quantity of
octahedral alum crystals to grow throughout the
liquid till it becomes a solid mass. It usually
happens that a trace of the salt either off the
stopper or the neck of the flask causes this crystallisation. With another solution, which is sulphate
of soda, the flask is tightly corked. On removing
the cork the air rushes in, the surface of the liquid is
slightly disturbed by the fall of a speck of dust, and
from this one single point crystals spread throughout the whole mass.

That it is not the entrance of air, but the
entrance of solid matter with the air which causes
crystallisation, can easily be seen with another
flask, which is simply plugged with cotton wool,
so that air obtains free access. On removing the
wool from the neck of the flask crystallisation
commences at some point on the surface of the
liquid. That the solid matter is dust floating in
the air may thus be shown: a flask of sulphate
of soda is provided with two tubes, one descend-
ing to the bottom of the liquid, the other just

passing through the cork ; these tubes are stopped
with cotton wool. Now on removing one of the
plugs and drawing air filtered by the wool through
the liquid no crystallisation takes place. If,
again, the air be taken from the lowest depths of
the lungs and blown through the liquid without
the intervention of the cotton wool there is still no
change ; but the air from a pair of bellows, ordinary
dusty air, at once causes the solidification to occur.
That the dust causing crystallisation is destructible
by heat is demonstrated in the following manner.
A cylinder of a supersaturated solution is carefully
uncovered, and the end of a glass rod which has
been heated, and afterwards cooled, is thrust in.
No change takes place, but the rod is now turned
round and the other end inserted, when the liquid
crystallises at once.   What is the nature of this
dust in the air destructible by heat which induces
crystallisation in this manner?

From the experiments of Mr. Liversidge,[1]
it seems more than probable, for it is the view
favoured by the majority of investigators of this
subject, that they consist of sulphate of soda.   It is
a salt which we know to be in the air, and solutions
of sulphate of soda, according to all experience, are
those which crystallise most readily on exposure.
Moreover, if precautions be taken to remove the

[1] 'Proc. Roy. Soc.' vol. xx. p. 497.

sulphate of soda from the air by means of baryta solutions, the remaining dust is powerless to cause crystallisation. Again, solutions of this salt, when uncovered in a very quiet room, the dust of which has settled for the most part, do not crystallise on shaking, but immediately on bringing a little powdered sulphate of soda into the room crystallisation takes place.

We therefore see that the sentence (p. 108), 'It would appear that specks of living matter may be born in suitable fluids, just as specks of crystalline matter may arise in other fluids. Both processes are really alike inexplicable,' is incorrect; crystallisation is as fully an intelligible process as any physical phenomenon ever investigated. The reason why crystals are formed in some places and under certain conditions is a matter undecided, though why they are formed at all is as evident as why water boils, or steam becomes water.

In order to believe that the cases are parallel, the imagination must be stretched to admit the idea that the tissues of the fungi said to have been evolved *de novo*, from di-sodic phosphate, and ammonium tartrate solution, consisted entirely of all or any of the four substances existing in such a liquid. 'Both crystals and organisms in such cases under suitable conditions' 'appear at first as minutest visible specks in solutions which were previously homogeneous.' ' In the one case we have to do with crystallisable matter in solution, and in the other with those big-atomed, unstable compounds which constitute the so-called *colloidal*[1] *state* of matter.' How are we to reconcile this statement with the evolution of life from some of the first order of crystalline substances, such as the solution of ammonium and sodium salts? Surely Dr. Bastian would not have us believe, in spite of all human knowledge, that these compounds are unstable and colloid substances? Yet this would seem to be the case, since he mentions the fact of cyanate of ammonia being converted by boiling into urea, and then states:[2] ' This would seem to show that the passage from the crystalloid to the colloid mode of molecular collocation is by

---

[1] With regard to the terms colloid and crystalloid, they are applied to solid substances soluble in water, which may be represented by gelatine on the one hand and common salt on the other. Colloids are not resolved into any definite crystalline form when assuming the solid state, whereas all the particles of a crystalloid substance are built on its own peculiar geometrical model, that of common salt being the cube, of alum the octohedron. A colloid can generally be resolved into a condition half-way between the solid and liquid states of matter, and is then commonly known as a jelly, but a crystalloid passes from the liquid to the solid suddenly and without any intermediate condition. A solution of a colloid is incapable of diffusion through a membranous material like bladder, while the reverse is the case with crystalloids, so that in this difference of properties we have a means of separating crystalloid from colloid substances.

[2] 'Nature,' vol. ii. p. 95.

no means a difficult one, that it may be brought about, in fact, by very slight determining causes.' As urea is well known to be one of the best crystallising substances in existence, this sentence is otherwise incomprehensible.

No one can say that there is any further resemblance between the growth of organisms and the formation of crystals, than that their origin is caused by similar bodies, and that under certain conditions they increase and multiply ; a shallow and useless comparison. A substance which is soluble in water is deposited when the water is removed by evaporation ; were organisms developed from solutions like crystals they would also be deposited from suitable liquids by evaporation, which is absurd. We must therefore have some statement, not only less vague but more definite, in fact a *definition* of what takes place, before we can listen to any such theory. As it stands now, too much is left to a very unscientific use of the imagination.

Before the recognition by Graham of these two classes of substances, there was a striking and inexplicable difference observable between the materials of inorganic and organic nature ; between lifeless and living matter ; the one generally being bounded by a hard and angular outline, the other by graceful and flowing curves. When we see how easy it is to mould colloid matter, how essen-

P

tially it is an elastic and a plastic material, how it
may be saturated with liquids which are not in a
condition of chemical combination, nor easily recognisable as mixtures, but constituting a state of
semi-solution, it is evident that colloid matter is
that upon which all forms of animal and vegetable
life depend for their existence ; they are, in fact,
built of colloid substances. Graham said that the
colloid was the dynamical condition of matter, and
the crystalloid the statical ; that is to say, that
colloid substances in solution were capable of un-
dergoing a change within themselves without the
exertion of external forces. For instance, a limpid
solution of silica will by itself become a mass of
jelly. Dr. Bastian has quoted Graham as the
mainstay of his arguments, but has reasoned most
illogically from his excellent observations. Organisms, according to him, are the molecular re-
arrangements of colloid substances in solutions ; he
sees a possibility of change from lifeless to living
matter as if he were unacquainted with the phenomena to which Graham refers. In the first place,
we know of no such re-arrangement of molecules
in colloid substances, otherwise we should have
many colloid compounds identical in composition,
but with different chemical properties, produced
spontaneously from each other. Then, as the
crystalloid is the statical condition of matter, it is

evident that it cannot of itself pass into the colloid state, and in the crystalloid state no such re-arrangement of molecules could take place. Diametrically in opposition to the view he advocates is his own statement that fungi and bacteria are produced from solutions of tartrate of ammonia, a substance of an eminently crystalline character.

The shiftiness of the arguments advanced by the defenders of heterogenesis is very striking. For instance, when liquids are found to be preserved through being protected by means of cotton wool, it is stated that the cotton may filter off, besides the living spores, the dead organic particles which it is inferred produce the change. When it is asserted that no change takes place in hermetically-sealed tubes, then it is by reason of the pressure of the confined gases that fermentative and putrefactive change is prevented. Next, on its being shown that fermentation and putrefaction go on under great pressures, it is because some fermentations are not prevented by pressure. When, again, the objection of pressure is removed from the experiments, the disruptive agency of heat is resorted to as an explanation. Having got rid of this difficulty, the next refuge is that sufficiently strong solutions were not employed. This argument failing, another is advanced, that the liquid was not an infusion of hay, and therefore not of the first order of fermentable matter (no experiments distinctly bearing on any one of all these statements being put in as evidence) ; and then as a last and least cogent reason, it is said that too low a power has been used in the microscopic examination of the liquids. The weakness of such vacillating reasoning is obvious.

Since the publication of my investigation, other workers have come into the field, whose evidence, more particularly from a physiologist's point of view, is of greater value than my own, such as that of the late Dr. Pode, of Oxford, and Prof. Ray Lankester, the well-known microscopist. Their mode of operating was in every respect exactly the same as that of Dr. Bastian, but they employed much higher magnifying powers ($\frac{1}{25}$ and $\frac{1}{30}$-inch object glasses) and proceeded more carefully ; for instance, in examining the liquids *before as well as after*, they were sealed in tubes. It is gratifying to find that their evidence is concordant with mine in every respect. From the results of 53 experiments made with turnip-infusion mixed with cheese, and hay-infusion, they produce ample evidence to prove the inaccuracy of Bastian's observations. He always supposes that having boiled his liquids for 10 minutes, he has really deprived them of living things ; but he never makes any attempt to prove this, and thus he begs the whole question. It is necessary that he should give unquestionable proof of his liquids being free from life when sealed in flasks, and such treatment as he subjects them to is in some cases, as Pasteur proved, totally inadequate for the purpose.

Dr. Cohn has studied this question attentively, and he finds that the extract of rennet, with which milk is curdled, contains an organised ferment which is chiefly a species of bacillus, probably identical with bacillus subtilis (the butyric ferment of Pasteur). This organism, after having been kept some time at a suitable temperature, develops a swelling at one extremity, which is filled with small round or oval particles, having the power of refracting light strongly.

The thread-like portion of the bacillus dies and forms a deposit in the liquid, but the rounded particles appear to be spores from which the complete organisms can be reproduced. Some of these are to be seen with short and very delicate threads attached as if they were germinating. It is quite in accordance with what we know of the Penicillium and other species of mould that these may be enclosed within the curd during the ordinary process of cheese-making, and be there protected from the injurious action of five or ten minutes boiling by the insignificant power of conducting heat which these substances possess.

Hence, fully developed bacteria are to be seen in closed flasks containing such a nutritive liquid and such solid particles. The probability of this was remarked (see p. 197) before the thorough research of Cohn had given us the knowledge requisite as the basis of a decided opinion.

Prof. Tyndall having very recently resumed work in connection with this subject, communicated his researches to the Royal Society on January 13, 1876. His investigation, which bears the title of the Optical Deportment of the Atmosphere with reference to the Phenomena of Putrefaction and Infection, has for its starting-point the discovery of Pasteur, that still air such as he found in the cellars of the Paris Observatory and on the Swiss glaciers may come in contact with previously heated infusions without these infusions generating or developing any life whatever, and that if solutions open to the air swarm with life it is because they have been impregnated with living particles floating in the air. Prof. Tyndall's previous researches in 1869 and 1870 have been already referred to (p. 174), and taking these in conjunction with later work it may now be accepted as an axiom that air which has lost its power of scattering light has also lost its power of producing life. A glass chamber filled with purified air remains dark even when placed in the track of a most powerful beam of light. Air can be rendered optically pure to this extent by simply leaving it undisturbed for two or three days. When the floating particles have subsided it will no longer transmit light, and solutions placed therein will remain unaltered though left for months, while similar solutions open to the ordinary air swarm with bacteria or some other organisms in twenty-four hours, or at most two or three days. An ingeniously contrived box standing on four legs, having a front of glass and two side windows, constituted a closed chamber of novel construction, the air of which could be examined with a ray of light as often as was felt desirable. Into the top of the box was fastened a piece of stout sheet india-rubber two inches in diameter, through the centre of which passed a funnel tube packed tightly with cotton-wool moistened with glycerine. To prevent the dislodgment of dust the inside of the box was painted with this fluid. The bottom of the box was pierced with two rows, sometimes with a single row of apertures, in which were fixed airtight, large test-tubes, intended to contain the liquid to be exposed to the action of the moteless air. The pliant india-rubber holding the funnel-tube admitted of its being moved so that its lower end could be brought over the mouths of the different test-tubes ;

liquids were thus easily poured in without disturbing the still air in the chamber. After a sufficient quantity of an infusion had been put into each tube it was boiled for five minutes by the tube being immersed in a bath of hot brine or oil, and the steam escaped from the chamber through tubes bent up and down about half-a-dozen times. During ebullition aqueous vapour rose from the liquid into the chamber, where it was for the most part condensed, the uncondensed portion escaping, at a low temperature, through the bent tubes at the top. Before the brine was removed little stoppers of cotton-wool were inserted in the bent tubes, lest the entrance of the air into the cooling chamber should at first be forcible enough to carry motes along with it. As soon as the temperature within the box had cooled down to nearly that of the surrounding air the cotton wool stoppers were removed.

The conditions to which the liquids were subjected are precisely those obtained by Pasteur in the open flasks with bent necks in which liquids were preserved after ebullition.[1] To quote Professor Tyndall's own words :—

'We have here the oxygen, nitrogen, carbonic

[1] See p. 166. In a lecture delivered at the London Institution in May 1872, the author showed experimentally that the air in such flasks was optically pure.

acid, ammonia, aqueous vapour, and all the other gaseous matters which mingle more or less with the air of a great city. We have them, moreover, "untortured" by calcination and unchanged even by filtration or manipulation of any kind. The question now before us is, can air thus retaining all its gaseous mixtures, but self-cleansed from mechanically suspended matter, produce putrefaction? To this question both the animal and vegetable worlds return a decided negative.

'Among vegetables experiments have been made with hay, turnips, tea, coffee, hops, repeated in various ways with both acid and alkaline infusions. Among animal substances are to be mentioned many experiments with urine; while beef, mutton, hare, rabbit, kidney, liver, fowl, pheasant, grouse, haddock, sole, salmon, cod, turbot, mullet, herring, whiting, eel, oyster have been all subjected to experiment.

'The result is that infusions of these substances exposed to the common air of the Royal Institution laboratory, maintained at a temperature of from 60° to 70° F., all fell into putrefaction in the course of from two to four days. No matter where the infusions were placed, they were infallibly smitten. The number of the tubes containing the infusions was multiplied till it reached six hundred, but not one of them escaped infection.

'In no single instance, on the other hand, did the air, which had been proved moteless by the searching beam, show itself to possess the least power of producing bacterial life or the associated phenomena of putrefaction. The power of developing such life in atmospheric air, and the power of scattering light, are thus proved to be indissolubly united.

'The sole condition necessary to cause these long-dormant infusions to swarm with active life is the access of the floating matter of the air. After having remained for four months as pellucid as distilled water, the opening of the back-door of the protecting case, and the consequent admission of the mote-laden air, suffice in three days to render the infusions putrid and full of life.

'That such life arises from mechanically suspended particles is thus reduced to ocular demonstration. Let us inquire a little more closely into the character of the particles which produce the life. Pour Eau de Cologne into water, a white precipitate renders the liquid milky. Or, imitating Brücke, dissolve clean gum mastic in alcohol, and drop it into water, the mastic is precipitated, and milkiness produced. If the solution be very strong the mastic separates in curds; but by gradually diluting the alcoholic solution we finally reach a point where the milkiness disappears, the liquid assuming, by reflected light, a bright cerulean hue. It is, in point of fact, the colour of the sky, and is due to a similar cause, namely, the scattering of light by particles, small in comparison to the size of the waves of light.

'When this liquid is examined by the highest microscopic power, it seems as uniform as distilled water. The mastic particles, though innumerable, entirely elude the microscope. At right angles to a luminous beam passing among the particles they discharge perfectly polarised light. The optical deportment of the floating matter of the air proves it to be composed, in part, of particles of this excessively minute character. When the track of a parallel beam in dusty air is looked at horizontally through a Nicol's prism, in a direction perpendicular to the beam, the longer diagonal of the prism being vertical, a considerable portion of the light from the finer matter is extinguished. The coarser motes, on the other hand, flash out with greater force, because of the increased darkness of the space around them. It is among the finest ultra-microscopic particles that the author shows the matter potential as regards the development of bacterial life is to be sought.

'But though they are beyond the reach of the microscope, the existence of these particles, foreign to the atmosphere but floating in it, is as certain as

if they could be felt between the fingers, or seen by
the naked eye, supposing them to augment in
magnitude until they come, not only within range
of the microscope, but within range of the unaided
senses. Let it be assumed that our knowledge of
them under these circumstances remains as defective
as it is now—that we do not know whether they
are germs, particles of dead organic dust, or particles
of mineral matter. Suppose a vessel (say a flower-
pot) to be at hand filled with nutritious earth, with
which we mix our unknown particles ; and that in
forty-eight hours subsequently buds and blades of
well-defined cresses and grasses appear above the
soil. Suppose the experiment when repeated over
and over again to yield the same unvarying result.
What would be our conclusion ? Should we regard
those living plants as the products of dead dust
or mineral particles ; or should we regard them as
the offspring of living seeds ? The reply is un-
avoidable. We should undoubtedly consider the
experiment with the flower-pot as clearing up our
pre-existing ignorance ; we should regard the fact
of their producing cresses and grasses as proof
positive that the particles sown in the earth of the
pot were the seeds of the plants which have grown
from them. It would be simply monstrous to
conclude that they had been " spontaneously generated."

'This reasoning applies word for word to the development of bacteria from that floating matter which the electric beam reveals in the air, and in the absence of which no bacterial life has been generated. There seems no flaw in this reasoning ; and it is so simple as to render it unlikely that the notion of bacterial life developed from dead dust can ever gain currency among the members of a great scientific profession.

'A novel mode of experiment has been here pursued, and it may be urged that the conditions laid down by other investigators in this field, which have led to different results, have not been strictly attended to. To secure accuracy in relation to these alleged results, the latest words of a writer on this question, who has influenced medical thought both in this country and in America, are quoted. "We know," he says, "that boiled turnip or hay infusions exposed to ordinary air, exposed to filtered air, to calcined air, or shut off altogether from contact with air, are more or less prone to swarm with bacteria and vibriones in the course of from two to six days." Who the "we" are who possess this knowledge is not stated.' Prof. Tyndall answers that he is certainly not among the number, though he has sought anxiously for knowledge of the kind.

The Professor next entered into details respecting the way in which he tested these and other similar statements of the writer alluded to, and said he felt bound to give them a most emphatic and unqualified denial. He thus continues :—

'The evidence furnished by this mass of experiments, that errors either of preparation or observation have been committed, is, it is submitted, very strong. But to err is human ; and in an inquiry so difficult and fraught with such momentous issues, it is not error, but the persistence in error by any of us, for dialectic ends, that is to be deprecated.'

Dr. Tyndall then shows by illustrations the risks of error run by himself. On Oct. 21 he opened the back-door of a case containing six test-tubes filled with an infusion of turnip which had remained perfectly clear for three weeks, while three days sufficed to crowd six similar tubes exposed to mote-laden air with bacteria. With a small pipette he took specimens from the pellucid tubes, and placed them under the microscope. One of them yielded a field of bacterial life, monstrous in its copiousness. For a long time he tried vainly to detect any source of error, and was perfectly prepared to abandon the unvarying inference from all the other experiments, and to accept the result as a clear exception to what had previously appeared to be a general law. The cause of his perplexity was finally traced to the tiniest speck of

an infusion containing bacteria, which had clung by
capillary attraction to the point of one of his
pipettes.

Again, three tubes containing infusions of
turnip, hay, and mutton, were boiled on Nov. 2
under a bell-jar containing air so carefully filtered
that the most searching examination by a concentrated beam failed to reveal a particle of floating
matter. At the present time every one of the
tubes is thick with mycelium and covered with
mould. Here surely we have a case of spontaneous
generation. Let us look to its history.

After the air has been expelled from a boiling
liquid it is difficult to continue the ebullition
without 'bumping.' The liquid remains still for
intervals, and then rises with sudden energy. It
did so in the case now under consideration, and
one of the tubes boiled over, the liquid over-
spreading the resinous surface in which the bell-
jar was imbedded, and on which, doubtless, germs
had fallen. For three weeks the infusions had
remained perfectly clear. At the end of this time,
with a view of renewing the air of the jar, it was
exhausted, and refilled by fresh air which had
passed through a plug of cotton-wool. As the air
entered, attention was attracted by two small spots
of penicillium resting on the liquid which had
boiled over. It was at once remarked that the

experiment was a dangerous one, as the entering air would probably detach some of the spores of the penicillium and diffuse them in the bell-jar. This was, therefore, filled very slowly, so as to render the disturbance a minimum. Next day, however, a tuft of mycelium was observed at the bottom of one of the three tubes, namely that containing the hay infusion. It has by this time grown so as to fill a large portion of the tube. For nearly a month longer the two tubes containing the turnip and mutton infusions maintained their transparency unimpaired. Late in December the mutton-infusion, which was in dangerous proximity to the outer mould, showed a tuft upon its surface. The beef infusion continued bright and clear for nearly a fortnight longer. The cold weather made a third gas-stove necessary in addition to the two which had previously warmed the room in which the experiments are conducted. The warmth of this stove played upon one side of the bell-jar; and on the day after the lighting of the stove, the beef-infusion gave birth to a tuft of mycelium. In this case the small spots of penicillium might have readily escaped attention; and had they done so we should have had three cases of 'spontaneous generation' far more striking than many that have been adduced.

From the earnestness with which Dr. Bastian

puts forth his conclusions, his experiments have received more attention than they would have gained on their own merits, for there is certainly nothing new, either in his mode of experimenting or in his results. Relying implicitly on their accuracy, he has revived an old theory and given it a new name. ' In a [1] not-living organisable fluid we have good reason to suppose that a living unit may originate ; and this being so we should have in such a case a veritable instance of the passage of the not-living into the living. Life would here begin *de novo* owing to the occurrence of certain new molecular combinations. To this process we propose to apply the name of *Archebiosis*.' In his communications to ' Nature,' and his two works, ' The Origin of the Lowest Organisms,' and ' The Beginnings of Life,' he makes unwarrantable assumptions and suggestions rather than distinct statements ; in fact, he argues from premisses which he has not first established. The evidence too which he quotes is not of the latest date, neither is it the product of exhaustive research, such as that of Pasteur.

He suggests or indicates the following arguments : 1st. That germ life if it exists in the air at all, ' is only very sparingly distributed.' [2] 2nd.

[1] ' Beginnings of Life,' vol. i. p. 232.
[2] Compare this with the account of Dr. Cunningham, Dr. Blackley, and Dr. Angus Smith's work.

That the particles in the air which produce fermentative changes are not germs. 3rd. That
those particles are fragments of dead nitrogenous
unstable organic matter. 4th. That all living
germs can be destroyed in liquids by boiling at
212° for ten minutes. 5th. That dead nitrogenous
unstable organic particles in contact with suitable
organic solutions can exert such an amount of
chemical action as to decompose the constituents
of the solution, recombine them in a different way,
and out of such newly formed material build living
organisms. Not a single point has he proved. The
burden of proof lies with him; those small ovoid
particles found floating by myriads in the air which
bear so close a resemblance to the spores of the
minute fungi that the·most able microscopists can
even assign names to them, must be shown by him to
be lifeless. They have, on the contrary, been shown
to be living, because when heated in water above
212° F. they ceased to have the power of reproduction in those solutions the fertility of which was
not affected by this temperature. The very theory
on which the supposed dead particles in an assumed unstable condition are dependent for their
power of communicating change to other matters,
is the useful but now long antiquated fiction of
catalysis. This theory originated with Berzelius,
and was adopted by Liebig to explain the pro-

cesses of fermentation and decay ; and concerning
it Graham wrote thirty-two years since.[1] 'It would
be unphilosophical to rest satisfied by referring
such phenomena to a force of the existence of
which we have no evidence. The doctrine of catalysis must be viewed in the light of a convenient
fiction, by which we are enabled to class together
a number of decompositions not provided for in
the theory of chemical affinity as at present understood, but which it is to be expected will receive
their explanations from new investigations.'

If on the one hand we are asked to believe the
arguments of Dr. Bastian, we cannot admit his
experiments, and on the other if we are asked to
believe his experiments, we cannot assent to his
arguments ; but as his reasoning and his experiments should both be in harmony to make out
a case, we see that he has failed completely in
doing this.   Again, if we compare both his reasoning and his experiments with those of M. Pasteur,
Prof. Lister, Mr. Ray Lankester, and Prof. Tyndall,
we can admit neither, and so find that he has
doubly failed.

Dr. Bastian apparently believes in a hypothetical substance producing hypothetical changes by
means of a hypothetical force.   His theory of

---

[1] 'Elements of Chemistry,' p. 197.

archebiosis, resting as it does on the fictitious catalytic action of supposed 'dead nitrogenous organic particles,' on so-called 'big-atomed unstable molecules' passing from the crystalloid to the colloid mode of molecular collocation is suggestive of the house which a man built upon the sand.

Further Researches on Atmospheric Dust.

Some idea of the varied nature of the dust floating around in the air, and which we are frequently conscious of swallowing in unpleasant quantities, may be given by just referring to the microscopic observations of several independent workers who have given attention to this matter. Dr. Angus Smith has shown that in this country, many miles from the sea, crystals of sea-salt are found in the rain ; they are, however, most abundant near the shore. Many years ago it was noticed that after a strong westerly gale the windows of Drayton Manor, near Tamworth, the residence of Sir Robert Peel, were covered with minute crystals of common salt, no doubt deposited by the evaporation of salt-water spray carried high into the air and across the country from the Welsh coast. The air of Manchester contains a considerable amount of sulphate of soda ; so, indeed, does that of all towns where much coal is burnt. It is found as a deposit, together with particles of iron rust, when rain water is evaporated.

Sulphate of soda, however, abounds most in the neighbourhood of chemical works or those towns which form the centres of chemical industry. When coal is burnt the sulphur is discharged into the air in a gaseous form, which ultimately becomes converted into sulphuric acid ; therefore the rain of all large towns is found to be acid, and it has a slow but distinctly destructive action upon such building stones as consist of magnesian limestone. Now if there be in the air traces of common salt, this in presence of the sulphuric acid would be converted into sulphate of soda ; hence the salt in country air would be converted into the sulphate of soda when it arrived at a large town. The air of mines, which has a peculiarly irritating action on the respiratory organs, is found to hold in suspension minute crystals of saltpetre and also probably of sulphate of potash. These crystals are derived from the gunpowder used in blasting, and at a time when the 'shots' or charges are fired it is blown into the air with the smoke. Blasting with gun-cotton has been found advantageous in some cases on this account, but the use of nitro-glycerine, on the other hand, has been the cause of severe headache, from a small quantity of the vapour of the substance being driven into the air, the inhalation

of which produces this effect on the workmen.
Dr. Angus Smith has examined the glittering
dust seen in the sunbeams during rapid railway
travelling, and found it to consist of minute rolled
plates of iron detached by friction from the rails
and wheels. During a railway journey near Bir-
mingham, Mr. Sidebotham remarked the gritty
nature of the dust in the carriage, and securing a
portion of it ascertained that 50 per cent. by weight
was composed of minute fragments of iron, some
particles being evidently the result of wear and
tear on the rails, while others were undoubtedly
cither the produce of neighbouring works or from
the furnace bars of the engine, as they were of the
nature of burnt iron and clinkers ; there were also
many small angular particles like cast-iron, having
a crystalline structure. Other portions of the dust
consisted largely of cinder, bits of coals, fragments
of yellow metal, opaque white and spherical bodies,
and bits of glass. It would be difficult to imagine
anything in the way of inorganic dust more irritating to the eyes, nose, and air passages. Mr.
Charles Stodder, of Boston, U.S., examined the
dust from a beam in the United States armoury
at Springfield, and found it to consist largely of
minute fragments of iron. To prevent such dust
circulating in the air, he recommended the use of
magnets fixed close to the grindstones and polish-

ing wheels in the workshops, a plan which was put in practice, and abandoned some years previously in this country.

An interesting communication was made by M. Nordenskjöld to the French Academy of Sciences in the summer of 1873. In December 1871 he observed that snow, collected at the end of a five or six days' continuous fall, was mingled with a large quantity of sooty-like dust consisting of an organic substance rich in carbon. It bore some resemblance to the meteoric dust which fell together with meteorites at Hessle near Upsal in the beginning of the year 1869. Suspecting the railways and houses of Stockholm of polluting the snow, he got his brother, who was living in a desert district in Finland, to give his attention to the matter, and the result was that he collected a similar powder. The snow gathered from floating ice in the Arctic regions and on the glaciers of Greenland, leaves, on melting, a greyish residue, consisting partly of the minute organisms known as diatomaceæ, but there are also particles of a considerable size containing iron, cobalt, nickel, silicon, carbon, and phosphorus.

The source of this dust is doubtful; two of its constituents, cobalt and nickel, are of rare occurrence in terrestrial matter, while on the other hand they are very commonly found in meteorites. It may be suggested that the dust may either be blown up from sandy plains situated at some distance from the place where it is deposited, or else discharged from the mouth of some far-off volcano. But what appears to be the best explanation of its origin is at the same time the most extraordinary. Prof. Nordenskjöld believed it to be meteoric dust dropped from the heavens.

Further Microscopic Examinations of Air disclosing the presence of Germs.

To examine ordinary air according to Dr. Angus Smith's plan, a wide-mouthed bottle, which may vary in size from four to sixty ounces, perfectly clean and dry, is filled with air of the desired place by means of the concertina bellows before mentioned, and into this is put about $\frac{1}{4}$ ounce of the purest possible distilled water, a commodity difficult to obtain free from floating particles. Very pure block ice perhaps yields it most easily. Without following the somewhat elaborate but undoubtedly successful plan of Mr. Cottrell, devised at the Royal Institution, one may take some carefully washed lumps of ice in a bottle and allow them to melt, pouring away the first two or three ounces of water after rinsing the glass. The bottle full of air to be examined is then shaken well, not to say violently, for a few minutes ; another quantity of air may be then introduced without a fresh addition of water, and the operation repeated as often as may be desirable. The liquid will soon become turbid, and the number of times the air has to be renewed to give visible turbidity varies very much with different localities. Mr. Stodder has raised objections to this plan of collecting dust from the air, but, as will be seen, it has yielded useful results.

In an experiment made by Dr. Smith with a bottle holding 5 litres, which was refilled 500 times with Manchester air, remarkable results were yielded by Mr. J. B. Dancer's examinations. The magnifying powers employed varied from 120 to 1,600 diameters. The bodies seen were—

1st. Particles of vegetable tissue, many of them partially burnt and quite brown in colour, exhibiting the pitted structure of fir-wood and other *Coniferæ*, probably wood used in lighting fires.

2nd. Fragments of vegetation resembling in structure hay, straw, and hay seeds.

3rd. Hairs of plants, and fibres resembling flax.

4th. Cotton-fibres, both white and coloured.

5th. Starch granules.

6th. Wool, white and coloured.

7th. In greatest abundance fungoid matter, spores and sporidia, varying in size from $\frac{1}{10,000}$ to $\frac{1}{50,000}$ of an inch in diameter.

Many of the spores were living and developed forms resembling rust or mildew. A calculation was made as to their number in the following manner.

Thus, under each field of the microscope there were more than 100 spores. In each drop of liquid there were over 250,000; the whole quantity consisting of 150 drops; there were then in this water *no fewer than 37½ millions of spores visible to a magnifying power of 1,600 diameters.* This quantity of air is the amount respired by an average-sized man actively employed during a period of 10 hours in Manchester.[1]

The researches[2] of Dr. Cunningham are of equal interest, and the most extensive on the subject yet published. They were made at Calcutta for sanitary purposes. By means of a vane attached to a tube a stream of air passed over a microscope slide moistened with glycerine, and the solid contents of the air were retained by this viscid liquid. The arrangement is really a weather-cock, which always presents the glass slide to the wind. The microscopic examination was made with powers varying from 400 to 1,000 diameters.

[1] 'Air and Rain, the Beginnings of a Chemical Climatology.' Dr. R. Angus Smith, F.R.S.

[2] 'Microscopic Examinations of Air.' D. Douglas Cunningham, M.B., Surgeon to H.M. Indian Medical Service. Published by Government, 1874.

At a height of about 5 ft. from the ground the following objects were collected :

1. Particles of siliceous matter.
2. Particles of carbonaceous matter.
3. Hair and animal substances.
4. Fragments of the cellular tissues of plants.
5. Pollen grain, amongst which were those of many common grasses, besides some of plants belonging to other natural orders.
6. Algæ. In comparison few in number. There were, however, undoubted fragments of *Oscillatoriæ*, *Desmidiaceæ Closterium*, and apparently *Diatomaceæ*, though these were very few.
7. Sporidiæ of lichens frequent.
8. Far the greater part of the bodies are sporidiæ of fungi, often referable at once to their proper genera.[1] Spores of *Macrosporium* and one or two other allied genera are extremely common. *Cladiosporium herbarum*, one of the most universally diffused fungi, appears in one case with a spore *in situ*. *Helminthosporium* is represented ; *Sporidesmium* is not unfrequent. The yeast fungus, a particular condition of common species of *Penicillium*, *Aspergillus*, and *Mucor* frequently occur, either in scattered particles or branched. A young *Mucor* with sporangia is seen amongst the other

[1] See Rev. M. J. Berkeley's article in 'Quarterly Journal of Microscopical Science,' April 1874.

objects drawn.   Spores of *Uredineæ Puccinia* and *Spheriaceæ*, with many others.

'The existence of *bacteria* in half of the specimen is also very worthy of consideration when the extreme rarity of such organisms in a recognisable form as a constituent of common atmospheric dust is recollected.'   The author concludes his work with the following remark :—' No connection can be traced between the numbers of *bacteria* spores, &c., present in the air and the occurrence of diarrhœa, dysentery, cholera, ague, or dengue, nor between the presence or abundance of any special form or forms of cells and the prevalence of any of these diseases.   Not less interesting are the researches of Mr. Blackley on *Catarrhus æstivus*, or hay fever.   They were made in the summer of 1873 during the months including April and July. Slips of glass covered with a liquid not likely to dry were exposed horizontally at a height of about 4 ft. 9 inches from the ground, and these were found to become covered with large quantities of pollen grains.   The greatest number obtained on a square centimètre of surface (⅜-inch square) in 24 hours, was 880, on June 28th.   Sudden diminutions in quantity were observed occasionally, and these were due to either a fall in the temperature or a shower of rain, or both causes combined.   A most ingenious method was employed to obtain some

knowledge as to the height at which this pollen
dust could be detected floating in the air. The
glass slips were attached to kites, and by fixing the
string of one kite to the body of another an elevation of 1,000 ft. was attained. By this means
the discovery was made that at the higher strata of
the air pollen was much more largely present than
at the 'breathing level,' in so large a proportion as
19 to 1. Abundant proof was obtained of the
presence of fungoid spores in large quantities in the
air. On one occasion the spores of a cryptogam at
the height of 1,000 ft. were found to be so numerous
that they could not be counted, but they were estimated to be not fewer in number than 40,000
to the square inch. That these minute organised
particles travel through the air to great distances
was proved by a series of experiments made near
Manchester, within the boundary of one of the most
populous parts and with no grass land within a
third of a mile. The quantity of pollen in this
locality was about $\frac{1}{10}$ that collected in the country.
Mr. Blackley considers he has proved that hay-fever
is caused by the inhalation of air containing pollen
in considerable quantity, and that the pollen adhering to the membranous lining of the larynx
and air passages and nasal lachrymal membrane
causes irritation, and the excessive secretion from
these parts. A solution of quinine, which is de-

structive to minute forms of life, has been shown by Helmholtz to be an effective application in cases of this disagreeable malady. The liquid is syringed into the nostrils. The Rev. W. H. Dallinger has shown that even the microscopic powers employed by Dr. Cunningham are insufficient, that it is only by employing object glasses of the highest powers, such as the $\frac{1}{25}$ and $\frac{1}{50}$, in conjunction with the most powerful eyepieces, that anything like the true number of organisms collected from the air can be seen. Dr. Lionel Beale, in his work on 'Disease Germs,' gives a drawing of *bacteria* germs magnified 5,000 diameters, and also minute fungus germs of much less size than the $\frac{1}{100,000}$ of an inch. If then these organised bodies are so extremely minute, how much more emphatic would Pasteur's evidence have been made had he been able in 1859 to have employed such enormous magnifying power ; how many more living corpuscles would he not have discovered ! It is a matter almost of surprise that he did so much with a microscope magnifying only 350 diameters.

In the 'Monthly Microscopical Journal' for August 1, 1873, there is an account of experiments by W. H. Dallinger and J. Drysdale, M.D., entitled 'Researches on the Life History of a Cercomonad ; a Lesson in Biogenesis.' It is a most instructive and suggestive paper, containing in a

very short space the results of a great deal of very
laborious research. They remark that 'the appearance or non-appearance of organic forms in
certain infusions placed in sealed flasks or tubes, or
otherwise conditioned, is held to be decisive of their
production *de novo* or otherwise, but in point of
fact we know *nothing*—absolutely nothing—of the
life history of the greater number of forms produced.' 'To attempt to decide therefore from the
experiments as yet published, that their production
in gross masses in inorganic infusions proves that
inorganic elements produced them, may be to beg
the whole question.' They speak of the advantage
secured by the combined investigation of two ob-
servers, and allude to the fact that objects have
been under constant observation for nine consecutive
hours, while the same drop of liquid has been
watched for three weeks. A peculiar flagellated
monad [1] noticed in the water in which the head of
a cod-fish was macerated, was seen to multiply by
fission for a period extending over from two to
eight days ; it then became amœboid. In this condition the mode of reproduction is changed ; two
individuals coalesce, slowly increase in size, and become a tightly distended cyst. This cyst bursts,
and an innumerable host of immeasurably small

[1] A somewhat oval-shaped organism to which is attached a long
whip-like appendage.

sporules are poured out as if in a viscid fluid and densely packed, barely recognisable with a $\frac{1}{50}$-inch objective magnifying 2,500 diameters. These sporules are scattered, they slowly enlarge, acquire a sort of motion like the fly-wheel of a watch, flagella are next formed, and they become active, attain rapidly the parent form, and once more multiply by fission. The average of forty observations on this fission shows that it occupied 4 minutes 40 seconds for completion. It was found that when adult forms and sporules together in a drop of liquid were evaporated to dryness and heated to 250° F., on moistening again all the adult organisms were found dead; but in two out of seven instances gelatinous points were seen like the early stage of developing sporules, and these after some hours' watching attained the flagellate state. Although in water at a temperature of 150° F. the adult monads are killed, *young monads appear and develope* in an infusion which has been raised to 260° F.; suggesting that the sporule is uninjured at a temperature much higher than that wholly destructive to the adult.

Cohn confirms the fact that Bacteria are not always killed by *boiling* in flasks, and that *Bacillus subtilis*, the lactic ferment, survives the boiling of the solution in which it is contained, and that in every case the boiling should be continued an hour. This establishes the accuracy of Pasteur's observations on the fermentation of milk.

Although we generally associate the growth of minute vegetation with the destruction of food, as, for instance, when bread or fruit becomes mouldy, and milk curdled and sour, yet in some cases the growth of the microscopic fungi effecting such changes converts plain viands into delicacies. For instance, the mellowing of a Stilton cheese is due to the spreading of a fungus, the *Aspergillus glaucus*, throughout the mass. The chemical change produced in the ageing of Roquefort cheese, which is owing to a like cause, has been made the subject of a careful investigation by a French chemist, M. Blondeau. The Roquefort cellars of Messrs. Rigal are teeming with the peculiar minute cryptogamic vegetation which we commonly call mould ; and into these places are the new cheeses brought to remain a year. At first they consist of pure caseine or curd free from fat, odourless and tasteless. After a year's time the cheese acquires a greyish-brown colour, a strong and peculiar odour, and piquant flavour. Subjoined in a tabular form are shown the constituent substances of the old and new cheese, the results of a great many analyses :—

R

Composition of New Cheese and Old Cheese kept a year.

New Cheese.	Per cent.	Old Cheese.	Per cent.
Caseine . . .	85·43	Caseine . . .	40·23
Fatty matter . .	1·85	Margarine . . .	16·85
Lactic acid . .	·88	Oleine . . .	1·48
Water . . .	11·84	Butyrate of ammonia .	5·62
	———	Caproate of ammonia .	7·31
	100·00	Caprylate of ammonia	4·18
		Caprate of ammonia .	4·21
		Common salt . .	4·45
Salt not estimated.		Water . . .	15·16
			———
			99·49

One-half the caseine by keeping becomes converted into other substances many of these being
compounds of ammonia with what are called fatty
acids.   These acids can be obtained by the oxidation of oleine, and are, in fact, produced when
butter becomes rancid.   The flavour of rancid butter
differs from that of Roquefort cheese, because the
acids in the latter substance are neutralised by
ammonia.   Caseine is a highly nitrogenous substance : the nitrogen it contains gets converted into
ammonia or some compound thereof, while portions
of its carbon and hydrogen are combined in such
a way as to form oleine.   This oleine undergoes
oxidation, producing the fatty acids which, instead
of being in the free state as in rancid butter, are
neutralised by the formation of ammonia salts.

It is more particularly to caprate of ammonia
that Roquefort cheese owes its peculiar sharp
flavour.

As was first shown by Cagniard de Latour and Schwann, the growth of a fungus at a prodigious rate, and the alterations it causes in the chemical composition of saccharine fluids, constitutes fermentation, the most essential process in all breweries and distilleries. In this case the yeast plant, a particular fungus of the torula form called *Saccharomyces cerevisiæ*, *Mycoderma cerevisiæ*, also *Hormiscium cerevisiæ*, is put into the liquid to be fermented. Its chemical action on the sugar is so extraordinary, and its own growth at the expense of the nitrogenous matter and phosphates in the wort so rapid, that were it not for microscopical research it would be impossible to believe the changes due to vital processes. The change consists in the conversion of sugar into alcohol and carbonic acid : there are other products, such as glycerine and succinic acid, but they occur only in small quantities. When a saccharine fluid containing nitrogenous matter and phosphates is exposed to the air it is always found to undergo a similar change.

The growth of other fungi often takes place in liquids which contain phosphates and ammonia salts or nitrogenous organic matter. Being excessively minute, these organisms require only exceedingly small quantities of matter for their nutrition, and as a large mass of evidence points to the conclusion that a most deadly class of diseases—for instance, most notably cholera—is due to the action of such organisms, it is of the highest importance that our drinking water should be so free from nitrogenous organic matter as to be incapable of affording sufficient nutriment to germ-life. Practical experience has shown that wherever water used for domestic purposes is contaminated even to a very small extent by organic matter of a nitrogenous nature (which is generally found to be derived from sewage), outbreaks of typhoid fever are common. Cholera has been distinctly traced to drinking water so contaminated, and the sewage spreading the mischief has in every case come from houses where the disease had been raging. It seems, however, that typhoid fever, which is always present more or less in this country, is propagated by germs present in all kinds of sewage matter, while the cholera germs are imported from Eastern countries and developed and transmitted in water to which they gain access. One of the most important applications of science to sanitary purposes is the chemical analysis of drinking-water, not that we can identify those special kinds of organic pollution in the water which are injurious, but we gain information as to whether it contains possibly hurtful matter or sufficient food for the nutrition of minute forms of life.

The Destruction of Organic Substances.

Broadly expressed, all nature consists of three divisions, the mineral, vegetable, and animal. Mineral matter in one form or another is the food and substance of plants. On the other hand, animals would starve on mineral food; the body is unable to convert it into substances which would replenish wasted tissues, neither has chemistry so far advanced that such a process could be devised. But animal life, in consuming vegetation while accomplishing its own nutrition, helps to return to earth and air, by the process of excretion, those simpler substances which form the sustenance of plants. From recently expressed views regarding the origin of life, the first living cell must have belonged to either a bacterium or one of the fungi, for it is from the supposed evolution of these forms of life from lifeless materials that deductions have been drawn. But neither bacteria nor fungi can exist on matter which, setting aside all life operations, is derived from the earth's crust and atmosphere, neither has it been attempted to be shown that they can be evolved from such materials. In truth, bacteria and fungi occupy a remarkable position, for while they may augment their substance by deriving nitrogen from ammonia salts or the ammonia of the air, carbon must be

presented to them in an unoxidised form, or perhaps, to speak more correctly, in an oxidisable form.

Animals and even plants themselves are by a
process of decay made into food for plant life.
Such a process as putrefaction or decay cannot be
without a cause, and such cause we cannot trace to
the very sluggish action of air or water at ordinary
temperatures. It is easy to obtain evidence
exactly corroborating the views of Pasteur published in 1863 :[1] 'Fermentation, putrefaction, and
slow combustion are the three natural phenomena
which concur in the grand operation of the destruction of organised matter, a necessary condition of
the perpetuity of life on the earth's surface.'

'Above all things, life, manifested in the production of the lowest organisms, appears to me one of
the essential conditions of these phenomena, but
life of a kind unknown until now, that is to say,
independent of air or free oxygen.'

'Dead matter which ferments or putrefies does
not yield, alone at least, to forces of a nature purely
physical or chemical.'

With regard to the statement concerning organisms capable of living without oxygen, in a
previous communication a particular fermentation
of tartrate of lime mixed with water had been
described, and an experiment of this kind was made.

[1] 'Comptes Rendus,' lvi. p. 734.

Tartrate of lime boiled with water to get rid of the dissolved oxygen and kill the germs was protected from the air by a layer of oil ; a minute quantity of the organised ferment, a species of vibrio, was then introduced under the oil, and this multiplied in the deposit of the lime salt. Similar facts were noticed in butyric fermentation and in the conversion of lactate into butyrate of lime. Another method of making the experiment consisted in filling a flask, having a curved tube-neck, with water and the tartrate ; the liquid was not boiled, but the tube-neck dipped into mercury to exclude air. *Monas* and *Bacterium termo* were developed in the liquid by means of their germs, and the oxygen dissolved in the water. In twenty-four to thirty-six hours all the oxygen was consumed to the last trace, and replaced by a somewhat smaller volume of carbonic acid : only then did the ferment, a species of vibrio, probably *Bacillus subtilis*, make its appearance. Strange to say, oxygen is not only unnecessary, but is poison to these living ferments. On passing air through the liquids containing them they sink to the bottom and the fermentation ceases. Carbonic acid, on the other hand, does not affect them. With regard to the part which oxygen plays in the destruction of organised matter, Pasteur discovered, by an analysis of the air in certain flasks (already alluded to) containing liquids, which had been

opened and resealed, that only a trifling absorption
of oxygen had taken place, notwithstanding the
duration of these experiments was three years, for
eighteen months of which time the temperature
was maintained at 86° F. The exact quantity of
oxygen absorbed was found in the carbonic acid
produced, allowing for the co-efficients of solubility
of the two gases in the liquids experimented on.
In the case of milk the oxidation of the fat had
caused the oxygen to disappear and be replaced
by carbonic acid ; it is usually so with other oily
matters. The liquids were otherwise unchanged.
being perfectly transparent as regards the sugar
solution and urine, while the milk was uncurdled
and possessed of its naturally alkaline reaction.
Sawdust from oak-wood boiled with a little water
absorbed from pure air only a few cubic centimètres
of oxygen in a month, while the same quantity not
heated and exposed to ordinary air absorbed nearly
140 cubic centimètres of oxygen in fourteen days,
and was found to be coated with mycelium fila-
ments and spores of *Mucedines*.

Perhaps of all matter the most putrescible is
blood. With the assistance of M. Claude Bernard,
Pasteur took specimens of arterial and venous blood
from a healthy dog, and kept it in flasks of pure air
at a temperature of 86° F. It never showed any
signs of putrefaction and retained its fresh smell.

After preservation for a month or six weeks the air of the flasks was analysed and the loss of oxygen was only 2 or 3 per cent.   In one case of a flask containing urine, the air was so little altered as to give on analysis the following percentage composition :—

Oxygen	19·2
Carbonic acid	0·8
Nitrogen	80·0
	100·0

This is a proof of the almost indestructible nature of organic substances of whatever kind when protected from the ravages of living matter.

Could anyone doubt the accuracy of these observations and conclusions, it is only necessary to turn to the recent very important experiments of Drs. Sanderson and Ferrier,[1] 'on the origin and distribution of microzymes (bacteria) in water.' They are of the greatest interest, and in addition to Pasteur's researches contribute greatly to our knowledge of what is constantly going on around us. No special method calling for description was employed.   To clean the vessels and tubes, after very careful washing they were heated as usual in such cases to 390° F., or thereabouts.   Certain of their sealed tubes containing liquids were heated to near

[1] 'Thirteenth Report of the Medical Officer of the Privy Council,' and 'Quarterly Journal of Microscopical Science,' vol. xi. p. 323.

that temperature, while, to protect liquids from the action of the air, plugs of cotton wool were used: the formation of microscopic fungi was thus prevented. In not a single instance did they find reason to suspect the evolution of organisms *de novo*, while they distinctly prove that bacteria arise from pre-existing germinal matter which has hitherto proved itself to be beyond the search of the microscope. This germinal matter is attached to all ordinary surfaces and is distributed in distilled water, being derived from contact with unclean vessels. Its presence was demonstrated in water which, both chemically and microscopically, seemed perfectly pure, and when tested by a beam of light the rays it scattered were of a blue colour, as is the case with the purest water.

Vessels washed and dried in the ordinary way with a towel contain this germinal matter. Now as it is impossible to admit of living matter being soluble in water, these living particles must be of excessive minuteness. Blood, urine, albumen, and Pasteur's sugar solution which never came in contact with any but previously heated vessels, when protected from atmospheric germs by plugs of cotton wool remained perfectly clear and limpid. When, however, a drop of cold distilled water was placed in the liquids, the contagion charged them with bacteria. One very remarkable experiment was the cutting of a piece of muscle from a freshly slaughtered rabbit by means of a previously heated knife, it being hung up by glass hooks, also cleansed by heat, under a bell jar. No putrefaction took place: a growth of mould simply formed on the surface.

Dr. Ferdinand Cohn has recently concluded that the microscopic power at his disposal is insufficient to yield him direct evidence of the existence of bacteria in the air, and he has made a set of experiments similar to those of Sanderson and Ferrier.

When air was passed through a solution capable of supporting and promoting the life and reproduction of bacteria, the fluid became turbid, and floating particles of mycelium appeared, which rapidly increased in size till they were easily recognisable. Aspergillus and penicillium were amongst the most commonly developed forms, while mucor only appeared once. The turbidity of the fluids was often due likewise to the presence of yeast cells, but it was remarkable that bacteria did not as a rule make their appearance. This is attributed by him not to the absence of bacteria germs in the air, of which we have abundant evidence, but to the circumstance that being extremely light and surrounded by a gelatinous envelope, they are not easily retained by the solution used, but are carried off by the air which escapes with each bursting bubble.   Cohn justly considers that germs cannot advance to a further state of development until they have been thoroughly moistened.

Regarding the action of bacteria on the liquids in which they live, the most important facts observed are : (1.) That their growth is attended with absorption of oxygen and discharge of carbonic acid.   (2.) That they are remarkably independent of the chemical constitution of the medium, provided they are supplied with oxygen ; and (3.) That they take nitrogen from almost any source which contains it and use it for building up their own protoplasm.

' It is this last power which specially indicates their place in nature as the universal destroyers of nitrogenous substances, acting as the pioneers if not the producers of putrefaction.' [1]

According to Pasteur, both bacteria and the organisms he calls vibriones co-operate in the process of putrefaction.

When a substance putrefies, for instance, a liquid, it is first deprived of all its dissolved oxygen by the bacteria.   They form a layer on the surface (zoogloea), the result being that no oxygen can penetrate to the liquid.   The vibriones are then developed, being protected from the too direct action

---

[1] ' Quarterly Journal Microscop. Soc.' pp. 326-7.   Cohn quotes and confirms these results.

of the air. The putrescible liquid then becomes the seat of two kinds of very distinct chemical actions related to the physiological functions of two different varieties of organisms which it nourishes. The vibriones, on the one part, living without the aid of atmospheric oxygen in the interior of the liquid, determine processes of fermentation, that is to say, they transform nitrogenous matters into more simple but nevertheless complex substances. The bacteria, on the other part, burn up those same products, and resolve them into the most simple binary compounds, water, ammonia, and carbonic acid.   Such are the results of putrefaction effected by means of free contact with the air. On the contrary, in the case of putrefaction out of contact with the atmosphere, the products of the splitting up of the putrescible matter remain unaltered. This explains why putrefaction in contact with air is a phenomenon which, if not always more rapid, is at least more complete, more destructive of organic matter than putrefaction out of contact with air. For example, if calcium lactate be putrefied out of contact with air, the vibrio ferment transforms the lactate into various products, amongst which calcium butyrate is always found. This new compound, undecomposable by the vibrio which has provoked its formation, will remain indefinitely in the liquid without any alteration. But repeat the

operation in contact with air, and it will be found
that in proportion as the vibrio ferment acts on
the inner portion of the liquid, the pellicle on the
surface consumes the butyrate gradually and completely. If the fermentation is very active, the
combustion at the surface is arrested, but only because the carbonic acid disengaged prevents the
access of atmospheric air. The phenomenon re-
commences as soon as the fermentation is completed
or slackened. This is, indeed the reason why, if a
natural saccharine liquid is fermented out of contact
with air, the liquid is charged with alcohol, whilst
if the operation takes place with access of air the
alcohol becomes acetified, burnt up, and transformed
entirely into water and carbonic acid : then the
vibriones appear, and following them putrefaction,
when the liquid contains no more than water and
nitrogenous matter.   Finally in their turn the
vibriones and the products of putrefaction are burnt
up by the bacteria, the last surviving of which cause
the destruction of those preceding them and thus
accomplish the complete return of organised matter
to the atmosphere and to the mineral kingdom.
Cohn's researches here again corroborate Pasteur's.

The advantage of this action on organic matter
being a life process, a function of living things, is,
that inorganic matter is not oxidised ; it is only that

kind of material which affords carbonic acid, water, and ammonia which is liable to attack.

The fetid smell evolved during putrefaction is due to certain sulphur compounds, the offensiveness increasing with the proportion of sulphur contained in the putrefying matter. The smell is scarcely perceptible, if the substance contains no sulphur. Every chemist is acquainted with the disgusting nature of most organic sulphur compounds and of sulphuretted hydrogen, one of the most constant products of the decomposition of albuminous matters, and therefore the bad odour of sulphurised products of putrefaction is easily intelligible.

In the consideration of solid substances under putrefaction Pasteur showed that, if all possibility of living germs gaining access to the interior of an animal were prevented, then after death putrefaction commenced on the surface. But if an animal be kept under usual conditions putrefaction commences in the intestinal canal, for there are found in this part, as Leuwenhoeck has noticed, fully developed vibriones. Being out of contact with air and bathed in liquid, their condition is that most favourable to multiplication and performance of their functions, and the body which has been preserved by the life and nutrition of its organs will then succumb to their action.

The living body is, as regards moisture, temperature, and material at hand, a most fitting place
for the development of life, from the multitudes of
germs that cannot escape being taken in with the
air breathed and the food consumed ; but the
natural functions interfere and the gastric juice, as
shown by Severi,[1] has the power of destroying the
agents of putrefaction and disease. As Dr. Lionel
Beale has pointed out, ' so long as the higher living
matter lives and grows, the vegetable germs are
passive and dormant, but when changes occur and
the normal condition departs, they become active
and multiply.'

Such are mainly the views of Pasteur, and it
must be admitted that they are the result of the
most extensive and elaborate research allied to the
most brilliant reasoning.

The liquid and solid constituents of meat react
on each other to produce to some extent a sort of
change which is entirely different from fermentation or putrefaction.   If the meat be enveloped in
a cloth steeped in alcohol and placed in a vessel so
that the vapours do not escape, there will be no
putrefaction, neither on the outside, because of the
antiseptic power of the alcohol vapour, nor within,
because vibriones are absent. If the quantity of
flesh thus preserved be small, it acquires a gamey

_____
[1] 'Zeitschrift für Chemie,' (2) iv. 285.

flavour, and if large it presents the appearance
of gangrene. There is no resemblance either in
nature or in origin between putrefaction and gangrene; indeed, gangrene may be likened to the
ripening of fruit after gathering.

As a necessary step to an exposition of the
changes that occur in fruit when ripening, it is
necessary to return to the action of ferments.
Ferments such as yeast are rendered less active
or produce a smaller amount of alcohol in exact
proportion to the quantity of free oxygen they
absorb. Under ordinary circumstances common
mould, when growing in a solution of sugar absorbs
oxygen from the air, feeds upon the organic matter
in the solution, and discharges carbonic acid without the production of any alcohol. If, however,
the mould be immersed in the liquid, and so
deprived of atmospheric oxygen, carbonic acid is
slowly liberated and alcohol produced, in other
words, the fungus under these altered conditions
acts as a ferment. In the former case, the free
oxygen absorbed from the air enters into combination with the carbon of the organic matter,
giving out heat, and yielding carbonic acid; it
resembles, in fact, ordinary respiratory combustion.
In the latter example, the life of the fungus continues by deriving from the decomposition of the
sugar the heat required for its existence. Alcohol

S

and carbonic acid are produced precisely in proportion to the duration of the acts of nutrition
of the plant under its changed existence. It is from
the small amount of oxygen in the fermentable
matter that the carbonic acid is derived. Evidently then under these latter conditions, in order
to produce the same amount of heat, or of carbonic
acid, a much greater quantity of saccharine matter
must be decomposed, much larger, in fact, than the
weight of the original organisms. These facts have
the following bearing upon the ripening of fruit.
When connection with the parent tree is severed
by, for instance, gathering, the cells are still living,
and a process similar to that of fermentation goes
on in the interior. From the saccharine matter and
the organic acids, alcohol and various ethers are
produced, giving the distinctive flavour to the fruit
and reducing its acidity. The exposure to the air
leads to an absorption of the oxygen ; the alcohol,
ethers, and sugar are changed to water and carbonic acid—this latter gas being equal in bulk to
the oxygen absorbed—the fruit loses flavour,
softens, and spoils, and is finally attacked by those
organisms whose life-processes are characteristic of
putrefaction. If, however, as Bérard has shown,
the fruit be immersed in carbonic acid, or some
other inactive gas, and so protected from oxygen,
a large amount of alcohol is produced. A very

perceptible amount of carbonic acid is still formed, but no softening of the fruit occurs. This is seen with bunches of grapes when taken out of the wine tub ; their flavour differs entirely from that of freshly gathered fruit, because they are immersed in carbonic acid. Fresh grapes enclosed in a vessel of this gas acquire the same flavour. Pasteur made an experiment with plums taken from a tree when nearly ripe ; twenty-four of these being placed in carbonic acid, yielded after a few days $6·5$ grammes of alcohol, while the fruit had not undergone any change, but was quite sound. A corresponding quantity of sugar had of course disappeared. The remaining twenty-four plums left in contact with air had become soft, watery, and very sweet. Grapes, melons, and all fruits containing acids behave thus. The living cells of the plums containing fermentable matters, act upon these as ferments do. Something similar to what occurs with the ripening of fruit may reasonably be considered as taking place with newly-mown grass ; a process of fermentation is commenced, developing a large amount of heat, and one of the products is the sweet scent of hay. Surely no better conditions for ripening fruit could be devised than those which exist inside a hay-stack, where, protected from oxygen, it would be surrounded by carbonic acid and kept warm ; and here theory accords with practice, for it is an old custom to ripen hard green pears by burying them an arm's length in a hay-rick.   By organised ferments are meant organisms which can directly assimilate oxygenated matters capable of supplying heat by their decomposition, such, for instance, as sugar.   Viewed in this light, fermentation appears to be a peculiar case of a very general phenomenon ; and all living things might be regarded as ferments in certain condi-tions of their lives, because none exist in which the action of free oxygen could not be momen-tarily suspended.   Now to generalise from this, let any living being be killed by asphyxia, by section of nerves, &c., or let any organ belonging to such a living being be deprived of connection with the neighbouring parts, the consequence is that since the chemical and physical operations of life cannot be instantly extinguished, they will continue, and if this happens with privation of oxygen, usually gaining access from within or without, then the being, the organ, or the cell will derive the heat required for its modified process of nutrition, or for change in its tissues, from the immediately surrounding materials.   From that moment it will decompose those materials, and the peculiar character of fermentation will be mani-fest if the quantity of heat developed corresponds to the decomposition of a weight of fermentable matter, perceptibly greater than the weight of materials usually set in action by the living being, by the organism, or by the cell. Pasteur's published researches have not yet been extended to the study of these details in animal organisms. It is probable that the phenomena may differ in certain particulars from those exhibited by vegetable cells. Doubtless further investigations by this great chemist will throw much light upon these obscure points so intimately connected with chemistry, physiology, and pathology, the phenomena of putrefaction and gangrene. However, as regards dead organic matter, whether of animal or vegetable origin, it would be well-nigh indestructible, if all those minute and apparently useless organisms which are the cause of putrefaction were themselves destroyed ;[1] 'life would be impossible, because the return to earth and air of all that had ceased to live would be suddenly suspended.'

[1] 'Comptes Rendus,' lvi. p. 738.   Pasteur.

LONDON : PRINTED BY
SPOTTISWOODE AND CO., NEW-STREET SQUARE
AND PARLIAMENT STREET